Unlocking the Secrets of Sleep

By
Clara Nightingale

Unlocking the Secrets of Sleep

Table of Contents

Introduction

Sleep is an essential and often overlooked cornerstone of our health and well-being. It's one of the most fundamental human needs, yet many of us struggle to achieve the restorative rest our bodies and minds crave. The journey to understanding and improving sleep quality begins with recognizing its undeniable impact on every aspect of our lives, from our physical health and emotional stability to our cognitive function and productivity.

In today's fast-paced, technology-driven world, the challenges to obtaining quality sleep are numerous. The prevalence of sleep disorders, increased stress levels, and the omnipresent glow of screens are just a few of the modern obstacles we face. However, with the right knowledge and strategies, we can reclaim our nights and, consequently, transform our days.

In this book, we aim to unravel the complex science behind sleep, providing you with a detailed yet approachable understanding of how sleep works. We'll explore the fundamentals of the sleep cycle, delve into the physiology and psychology of sleep, and highlight the myriad health benefits that come from getting a good night's rest. Whether you're dealing with insomnia, stress-related sleep issues, or just looking to optimize your sleep quality, this comprehensive guide is designed to address all your concerns.

It's not just about the quantity of sleep but the quality, too. We'll guide you through practical steps for improving sleep hygiene, incorporating nutrition and exercise for better rest, and managing stress effectively. You'll also learn about the latest technology in sleep tracking and how to adapt your habits to make the most of these advancements.

By the end of this journey, you'll possess a richer understanding of sleep and be equipped with a toolkit of strategies to enhance this vital part of your life. Let's embark on this transformative path towards better sleep and, ultimately, better health and happiness.

Chapter 1:
The Fundamentals of Sleep

Understanding the fundamentals of sleep is essential for anyone aiming to improve their sleep quality and overall well-being. At its core, sleep is a complex and dynamic process that involves various stages and cycles, each playing a critical role in maintaining our physical and mental health. These stages of sleep, from light to deep sleep and the REM phase, follow a predictable pattern known as the sleep cycle. Coupled with the sleep cycle is the circadian rhythm, our internal clock that governs the timing of sleep and wakefulness. Together, these elements form the architecture of sleep, orchestrating a delicate balance vital for restorative rest. By delving into the basics of how sleep works, we can begin to appreciate its impact on our daily lives and take the first steps toward healthier sleep practices.

Understanding the Sleep Cycle

Let's dive into one of the most fascinating aspects of sleep: the sleep cycle. Understanding how sleep unfolds in a series of stages can give you the tools to enhance your own sleep quality and well-being. Essentially, the sleep cycle is a repeating pattern that our bodies go through several times during the night, each one lasting about 90 minutes.

This cyclical nature is divided into two main categories: Non-REM (NREM) sleep and REM (Rapid Eye Movement) sleep. NREM sleep consists of three stages, each progressively deeper than the last.

These stages collectively form the foundation for restorative sleep, helping the body recover from the day's activities and preparing it for the next. While in NREM sleep, your brain waves slow down, your heart rate decreases, and your body temperature drops.

REM sleep, on the other hand, is often dubbed the "dreaming stage." During REM, your brain activity ramps up, becoming almost as active as when you're awake. This is when vivid dreams occur, and it plays a crucial role in cognitive functions, such as memory consolidation and emotional processing.

Intriguingly, the balance between these sleep stages changes throughout the night. In the earlier cycles, we spend more time in the deeper stages of NREM sleep, crucial for physical restoration. As the night progresses, the REM stages become longer. This fluid transition between stages is a critical aspect of what makes sleep so rejuvenating.

Disruptions to this intricate cycle, whether from stress, poor sleep hygiene, or other factors, can severely impact your overall sleep quality. When the cycle is harmonious, you wake up feeling refreshed. But when it's not, mornings can be a struggle.

By becoming aware of your sleep cycle, you can make informed decisions to improve your sleep environment, routines, and lifestyle choices. Little adjustments can lead to significant differences in how rested you feel and how efficiently your body functions. Understanding these cycles isn't just for curiosity—it's a stepping stone towards better health and sharper cognition.

Stages of Sleep come under the umbrella of the 'Understanding the Sleep Cycle' section, and it's a fundamental cornerstone for anyone seeking better sleep and well-being. To appreciate the stages of sleep, imagine they're akin to an orchestra: Each stage plays a distinct role in creating a harmonious and restorative symphony. There are four primary stages, each with its unique characteristics and contributions to our overall sleep quality.

The first stage, known as NREM Stage 1, serves as the gateway to sleep. It's a light sleep phase that typically lasts just a few minutes, where you drift in and out of consciousness and can be easily awakened. This stage is characterized by the slowing down of brainwaves, specifically the transition from alpha waves (which dominate when you're awake) to theta waves. Think of it as the prelude of the night's rest, easing you into the sleep cycle and preparing your body for the deeper stages of sleep to come.

Following this, we proceed to NREM Stage 2. This stage isn't just longer but more substantial, accounting for approximately half of the total sleep cycle. At this juncture, there's a further slowing of brain activity, interspersed with bursts of rapid brain waves known as sleep spindles. These spindles are essential for memory consolidation and cognitive function. Additionally, features known as K-complexes appear, which serve as markers that help you stay asleep while still being able to respond to the external environment. During NREM Stage 2, your body's temperature drops, and both your heart rate and breathing normalize, making it a critical phase for physical restoration.

NREM Stages 3 and 4 together comprise what is often referred to as deep sleep or slow-wave sleep (SWS), with Stage 3 acting as a transitional period to Stage 4. It's in these stages that the most profound and restorative processes occur. Characterized by delta waves, the slowest and highest amplitude brainwaves, this stage is when the body undergoes significant repair. Tissue growth and muscle repair are paramount here, and the body releases growth hormones that play a vital role in physical recovery. Deep sleep also influences aspects of memory and learning, specifically consolidating declarative memories, such as facts and knowledge-based information.

Following the deep sleep stages, we transition into the REM (Rapid Eye Movement) stage, a phase that captures much of the intrigue around sleep. Lasting roughly 90 minutes each cycle and getting progressively longer with each successive cycle, REM sleep is a captivating blend of heightened brain activity and muscle atonia (near paralysis). It's paradoxical in nature; your brain is quite active while your body remains largely immobile. REM sleep is the stage most closely associated with vivid dreaming and plays a foundational role in emotional regulation and memory consolidation, particularly procedural memory and skills learning.

Interestingly, the sleep cycle doesn't progress sequentially through the stages but instead oscillates. A typical night's sleep involves cycling through NREM and REM stages approximately four to six times, with each cycle lasting about 90 to 120 minutes. Early in the night, NREM stages dominate, particularly deep sleep, acting as the body's heavy-lifters in physical and mental repair. As the night progresses, REM stages become more prolonged, emphasizing cognitive and emotional restoration.

The delicately balanced dance of these stages exemplifies the intricate architecture of sleep. Any disruption to this cycle, whether from lifestyle choices, stress, or sleep disorders, can profoundly impact overall sleep quality and health. Understanding and respecting these stages can be a stepping stone toward better sleep hygiene and overall wellness.

Moreover, the effects of these sleep stages extend beyond just nightly rest. Long-term disruptions in these stages are linked to a myriad of health complications, underscoring the importance of a balanced sleep architecture. From cardiovascular health issues to cognitive impairments and mood disorders, the ramifications of not achieving optimal sleep stages are vast and interconnected.

For those struggling with achieving or maintaining deep sleep or REM sleep, it's essential to recognize that while the body has a remarkable ability to adapt, persistent disruptions can accrue a sleep debt that's hard to repay fully. Practicing good sleep hygiene, managing stress, and creating a tranquil sleep environment can be crucial steps towards optimizing these sleep stages. Equally important is seeking medical advice if sleep disorders are suspected, as professional intervention can provide specific recommendations and treatments tailored to individual needs.

In summary, the stages of sleep are more than just progressions through the night—they are meticulously timed processes that ensure our bodies and minds are rejuvenated and prepared for the next day. Being conscious of how these stages work and the importance they hold in our overall health can be empowering. By enhancing our understanding, we can take actionable steps to improve our sleep quality, leading to a more vibrant and healthier life.

The Role of Circadian Rhythms is a fundamental concept that significantly impacts how well you sleep and your overall well-being. Understanding your body's internal clock can be a game-changer in improving sleep quality. Essentially, circadian rhythms are 24-hour cycles that regulate various physiological processes, including the sleep-wake cycle. These rhythms are influenced by external cues such as light and darkness, making them deeply intertwined with nature.

Circadian rhythms are primarily governed by a group of neurons in the hypothalamus known as the suprachiasmatic nucleus (SCN). This internal clock responds to light cues from the environment, signaling your body when it's time to be awake and when to rest. When your rhythms align well with the natural day-night cycle, you experience better sleep quality, increased alertness during the day, and overall improved health. Conversely, disruptions in these rhythms can lead to poor sleep, fatigue, and various health issues.

Exposure to natural light is one of the most potent factors that influence circadian rhythms. When you wake up in the morning, sunlight entering your eyes triggers a signal to the SCN, which then halts the production of melatonin, the hormone responsible for making you feel sleepy. This process helps you feel more awake and alert. As the day progresses and natural light diminishes, melatonin production ramps up, preparing your body for sleep. Therefore, maintaining regular exposure to natural light is crucial for keeping your circadian rhythms in sync.

Artificial light, particularly blue light emitted from screens, can significantly disturb your circadian rhythms. Blue light has a wavelength that mimics daylight, tricking your brain into thinking it's still daytime, even when it's late at night. This delays melatonin production and can make it harder for you to fall asleep. To counteract this, it's advisable to limit screen time at least an hour before bed and consider using blue light filters on your devices.

Shift work and jet lag are common scenarios where circadian rhythms can become misaligned. If you work night shifts or frequently travel across time zones, your internal clock can struggle to adapt to the new schedule. This misalignment results in irregular sleep patterns, diminished sleep quality, and increased susceptibility to various health problems. Strategies such as timed light exposure, melatonin supplements, and gradual adjustments to your sleep schedule can help mitigate these effects.

Another important aspect to consider is the role of lifestyle habits in maintaining healthy circadian rhythms. Regular exercise and a balanced diet can support your internal clock. Exercise not only promotes physical health but also helps regulate hormones and body temperature, both of which are essential for maintaining a stable circadian rhythm. Eating meals at consistent times also reinforces your body's internal clock, signaling when it's time to be awake and when to rest.

Stress and emotional well-being also interact with your circadian rhythms. High levels of stress can lead to increased levels of cortisol, a hormone that can disrupt your sleep-wake cycle. Mindfulness practices, such as meditation and deep breathing exercises, can help manage stress levels and thus promote healthier circadian rhythms. Creating a routine that includes relaxation techniques can be incredibly beneficial for aligning your internal clock.

It's important to note that circadian rhythms do not remain static throughout life. They evolve as you age, adapting to changes in lifestyle and environmental factors. For instance, teenagers naturally tend to have a delayed sleep phase, meaning they feel alert later in the evening and find it challenging to wake up early. Understanding these shifts can help tailor sleep strategies to different stages of life, ensuring that your circadian rhythms remain balanced regardless of age.

Finally, people with certain sleep disorders often have disrupted circadian rhythms. Conditions like delayed sleep phase syndrome (DSPS) or advanced sleep phase syndrome (ASPS) result from a misalignment between the internal clock and the external world. These disorders are not merely inconveniences but can significantly affect daily functioning and quality of life. Treatments often involve a combination of light therapy, melatonin supplements, and behavioral interventions to realign circadian rhythms with the desired sleep-wake schedule.

In summary, circadian rhythms play a pivotal role in determining your sleep quality and overall health. Aligning your internal clock with natural light and maintaining consistent daily habits are essential steps for enhancing sleep and well-being. Whether you're dealing with shift work, managing stress, or simply looking to improve your nightly rest, understanding and respecting your circadian rhythms can provide a roadmap to better sleep and a healthier life.

Sleep Architecture

When we lay our heads down at night, our brains don't simply switch off; they embark on a highly organized journey through distinct stages. Sleep architecture refers to the intricate structure of this journey, covering the various stages and their cyclical pattern throughout the night. It's like a rhythmic dance, under the guidance of our inner biological clocks.

Each night, we typically cycle through four to six sleep cycles, each lasting about 90 to 110 minutes. These cycles are composed of non-rapid eye movement (NREM) and rapid eye movement (REM) sleep. NREM sleep is further divided into three stages: N1, N2, and N3. The N1 stage is light sleep, a transition phase between wakefulness and deeper sleep. During the N2 stage, brain waves slow down, eye movements cease, and body temperature drops. Finally, in the N3 stage, also known as slow-wave or deep sleep, the body repairs itself, and the immune system strengthens.

Following NREM, we enter REM sleep, where vivid dreams usually occur. During REM, brain activity ramps up almost similar to wakefulness, but the body remains in a state of temporary paralysis—an evolutionary feature thought to prevent us from acting out our dreams. This phase is crucial for cognitive functions like memory consolidation, emotional regulation, and learning.

One fascinating aspect of sleep architecture is how it shifts as the night progresses. In the earlier part of the night, N3 deep sleep dominates, ensuring the body gets its required physical restoration. As morning approaches, REM sleep stages elongate, benefiting mental and emotional well-being. This architecture is finely tuned to harmonize both physical recovery and mental processing.

Several factors can influence sleep architecture, altering the balance between NREM and REM sleep. For instance, age plays a significant role; children can spend up to 50% of their sleep time in REM, compared with about 20-25% in adults. Stress, medication, and health conditions such as sleep apnea or insomnia can also disrupt the natural rhythm of sleep stages, sometimes reducing the amount of restorative deep sleep or REM sleep one receives.

Recent studies have identified the role of circadian rhythms in sleep architecture. These 24-hour cycles, governed by the brain's suprachiasmatic nucleus, influence the timing and duration of sleep stages. However, disruptions to these rhythms—like irregular sleep schedules or exposure to artificial light—can impair the architecture of sleep. This is why maintaining consistent sleep patterns and managing light exposure is advice often given to those struggling with sleep issues.

Understanding sleep architecture helps us appreciate the complexity and necessity of each sleep stage. It's not just about getting eight hours of sleep but about cycling through these stages efficiently. Disruptions in this architecture can lead to grogginess, poor cognitive function, and a number of health issues over time. Thus, aiming for not just quantity but quality of sleep becomes paramount in achieving overall well-being.

Chapter 2:
The Physiology of Sleep

As we transition from understanding the fundamentals of sleep, it's crucial to delve into the physiological mechanisms that orchestrate this vital function. Sleep is a complex, highly regulated process driven by both the brain and the body. The brain houses the master clock in the suprachiasmatic nucleus, which synchronizes our circadian rhythms with the external environment. This intricate dance involves the pineal gland releasing melatonin in response to darkness, signaling that it's time to rest. Meanwhile, neurotransmitters like GABA and adenosine work behind the scenes to decrease neural activity and promote sleep onset. REM sleep, often dubbed the "dream stage," is essential for cognitive functions such as memory consolidation and emotional regulation. For those striving to improve their sleep quality and overall well-being, understanding these physiological underpinnings provides a foundation on which to build healthier sleep habits.

How the Brain Regulates Sleep

Our brain plays an intricate role in managing when we fall asleep, how deeply we sleep, and when we wake up. Central to this process is the brain's relationship with our internal biological clock, scientifically known as the circadian rhythm. This internal clock is mainly controlled by the suprachiasmatic nucleus (SCN), a small region in the hypothalamus. The SCN receives light signals from the retina and uses this information to orchestrate various hormonal and neural responses that dictate our sleep-wake cycle.

Throughout the day, the brain orchestrates a symphony of neurochemical activities to prepare us for sleep. As evening approaches, the SCN signals the pineal gland to secrete melatonin, a hormone that induces drowsiness. This is the body's way of saying it's time to wind down. As melatonin levels rise, body temperature drops slightly, and our body begins to transition to a state suitable for sleep.

Deep within the brain, numerous neurotransmitters play significant roles in regulating the sleep cycle. Gamma-aminobutyric acid (GABA) is one such inhibitory neurotransmitter that helps calm brain activity, leading to the onset of sleep. It works like a braking system, ensuring that our minds can shift from the alertness of daytime to the calm needed for sleep. Conversely, neurotransmitters like norepinephrine and dopamine are reduced to help keep the brain quiet during bedtime.

Another critical player is the hypothalamus, which governs various aspects of sleep regulation. Collaborating with the brainstem, the hypothalamus ensures the smooth transition between different stages of sleep. These structures work in unison, also involving the thalamus, to modulate sensory input and create the peaceful environment our brain needs to rest and rejuvenate.

Understanding how the brain regulates sleep can empower us to take practical steps toward better sleep hygiene. By aligning our daily habits with our brain's natural sleep-regulating processes, we can enhance our overall sleep quality and well-being. Recognizing the roles of melatonin, neurotransmitters, and the hypothalamus offers insight into creating a lifestyle conducive to restful and rejuvenating sleep.

The Pineal Gland and Melatonin are key players in the intricate dance of sleep, subtly orchestrating our body's nightly rhythms.

Nestled deep in the brain, the pineal gland is tiny—no larger than a grain of rice—but its impact on sleep is profound. This minuscule endocrine organ produces melatonin, a hormone that signals to your body when it's time to rest.

The pineal gland's role in sleep begins as the day draws to a close. As darkness envelops your environment, your eyes perceive the diminishing light through specialized cells in the retina. This information is relayed to the suprachiasmatic nucleus (SCN) in the brain, often thought of as the body's master clock. The SCN then sends signals to the pineal gland, triggering the release of melatonin. With melatonin levels rising, your body starts to wind down, preparing itself for the restorative processes of sleep.

Melatonin is often dubbed the "sleep hormone," but its role extends beyond merely inducing drowsiness. It helps regulate your circadian rhythms—those 24-hour cycles that govern not only your sleep-wake patterns but also numerous physiological processes, such as hormone release and body temperature regulation. By aligning these rhythms with the external environment, melatonin helps ensure that you fall asleep and wake up at appropriate times.

While most people produce adequate amounts of melatonin naturally, several factors can disrupt this delicate balance. Exposure to artificial light, particularly blue light emitted by screens, can mimic daylight and trick the pineal gland into reducing melatonin production. That's why it's crucial to limit screen time in the hour or two before bed, creating a dark and calming environment to signal to your body that it's time to sleep.

Interestingly, melatonin production is also influenced by age. Children and adolescents tend to have higher levels, which can explain why they often need more sleep. As we age, the pineal gland's ability to produce melatonin wanes, making it more challenging for older adults to maintain consistent sleep patterns. This decline in melatonin production is a natural part of aging, but it can exacerbate sleep difficulties if not addressed through good sleep hygiene practices and, in some cases, supplementation.

It's important to understand that melatonin is not a sedative; it doesn't knock you out like some sleep medications. Instead, it works by nudging your internal clock in the right direction, making it easier for you to transition into sleep. For those who struggle with jet lag or shift work, melatonin supplements can be a helpful tool to reset their circadian rhythms. However, supplementation should be approached cautiously and preferably under the guidance of a healthcare professional because overuse can disrupt your natural hormone production and worsen sleep problems in the long run.

The pineal gland's function and the role of melatonin are areas of ongoing research. Scientists are continually seeking to understand the myriad factors that can impair melatonin production, exploring not just the impact of light exposure but also diet, stress, and even genetic factors. These insights are crucial in developing strategies to optimize sleep quality, particularly for those suffering from chronic sleep issues like insomnia.

Given the pineal gland's sensitivity to light, adopting practices that promote darkness in the evening can significantly enhance your sleep quality. Simple changes, such as dimming lights an hour before bedtime, using blackout curtains, and avoiding electronic devices, can encourage your pineal gland to produce the melatonin you need for a good night's sleep. Sometimes, even small tweaks to your environment and daily routines can yield substantial improvements.

When melatonin levels are well-regulated, the benefits extend beyond just sleep. Proper melatonin production contributes to overall well-being, affecting mood, immune function, and even cardiovascular health. As such, prioritizing practices that support your pineal gland's function can have far-reaching effects on your overall health. Your body's internal clock is a finely tuned system, and melatonin is one of its most crucial gears.

In summary, the pineal gland and melatonin play an indispensable role in governing your sleep-wake cycles. By understanding and supporting this natural process, you can enhance not only your sleep quality but also your overall health and well-being. It's a delicate balance, but with a bit of knowledge and effort, you can optimize your body's natural rhythms and enjoy the restorative sleep you deserve.

Neurotransmitters and Sleep are essential components in the delicate balance and regulation of our sleep cycles. Our brain uses neurotransmitters as chemical messengers to communicate between neurons, and many of these play a crucial role in promoting both sleep and wakefulness. Understanding how these neurotransmitters function gives us insight into why we sleep the way we do and how disruptions in their activity can lead to sleep disorders.

One of the primary neurotransmitters involved in sleep is serotonin. Found predominantly in the brainstem, serotonin is a precursor to melatonin, the hormone that governs our sleep-wake cycle. Serotonin levels typically rise and fall throughout the day, peaking during daylight hours to promote alertness and gradually declining as evening approaches, facilitating a state of relaxation and readiness for sleep. When serotonin's balance get disrupted, it can manifest in various sleep issues, including insomnia and difficulties in maintaining restful sleep.

Norepinephrine, another key neurotransmitter, is associated with alertness and wakefulness. Primarily synthesized in the locus coeruleus, norepinephrine's role is to prepare the body for action by increasing heart rate and blood flow to muscles. During REM sleep, its levels drop significantly, allowing the body to relax and rejuvenate. However, an overactive norepinephrine system, often linked to stress or anxiety, can make it challenging to fall or stay asleep, contributing to fragmented sleep patterns.

GABA (gamma-aminobutyric acid) serves as the brain's primary inhibitory neurotransmitter, calming the central nervous system and promoting sleep. By reducing neuronal excitability, GABA allows the brain to transition from wakefulness to sleep. Many sleep medications, like benzodiazepines, work by enhancing GABA's effects, acting as potent sleep aids. Without sufficient GABA activity, the brain can become too stimulated, making it hard to unwind and sleep soundly.

Dopamine, often associated with pleasure and reward, also plays a role in regulating wakefulness. Emerging research indicates that dopamine levels follow a circadian rhythm, peaking in the morning to stimulate wakefulness and energize starting the day. Disruptions in dopamine pathways, as seen in conditions like Restless Leg Syndrome (RLS) or Parkinson's disease, often accompany significant sleep disturbances, further illustrating the interplay between neurotransmission and sleep health.

Acetylcholine, crucial for rapid eye movement (REM) sleep and facilitating dreams, remains another pivotal neurotransmitter. Its activity peaks during REM sleep, supporting cognitive processes and memory consolidation. Reduced acetylcholine function can lead to insufficient REM sleep, impacting emotional health and cognitive function. Cholinergic activity, thus, ensures a balanced sleep cycle, highlighting its importance in overall sleep quality.

Additionally, histamine is a neurotransmitter that promotes wakefulness and is part of the brain's arousal system. Histamine-producing neurons activate in the morning, sustaining alertness throughout the day. Antihistamines, commonly used as over-the-counter sleep aids, work by blocking histamine receptors, thereby inducing drowsiness. Yet, overuse can lead to dependency and potentially disrupt normal sleep architecture.

Another essential compound, adenosine, accumulates in the brain throughout wakefulness and creates pressure to sleep. As adenosine levels increase, we start to feel tired, signaling it's time to rest. Caffeine works by blocking adenosine receptors, temporarily staving off sleepiness. However, this can lead to a build-up of adenosine, culminating in more profound fatigue once the caffeine wears off.

In summary, the balanced interplay of various neurotransmitters—serotonin, norepinephrine, GABA, dopamine, acetylcholine, histamine, and adenosine—is fundamental to regulating sleep. Each one uniquely influences different aspects of our sleep-wake cycle, working together to ensure that we transition smoothly from wakefulness to sleep and back again. Disruptions in any of these neurotransmitter systems can result in sleep disorders, impacting overall health and well-being. Understanding their roles and interactions can empower us to take more informed steps in improving our sleep quality.

The Importance of REM Sleep

Rapid Eye Movement (REM) sleep is a vital component of the sleep cycle, offering benefits that extend well beyond mere rest and rejuvenation. It's the stage where the brain is most active, allowing for critical processes such as memory consolidation and emotional regulation. REM sleep is fascinating because it's when the mind engages in vivid dreaming, providing a unique window into our subconscious thoughts and fears.

Unlike the other stages of sleep, REM is characterized by brain activity that closely resembles wakefulness. The brain cycles through different stages of sleep approximately every 90 minutes, but it's during REM that most dreaming occurs. During this time, the brain consolidates new information, converting short-term memories into long-term ones. This consolidation process is crucial for learning and cognitive function, making REM sleep indispensable for students, professionals, and anyone keen on optimizing mental performance.

Another important aspect of REM sleep is its role in emotional regulation. Studies have shown that the brain processes emotional experiences during this stage, helping individuals cope with stress and trauma. Imagine REM sleep as a therapeutic session where your unconscious mind sorts through the emotional baggage you've accumulated throughout the day. It allows you to wake up feeling more balanced and emotionally resilient. This is especially crucial for insomniacs, who often experience heightened levels of stress and anxiety. Quality REM sleep can provide a natural buffer against these emotional challenges.

Furthermore, REM sleep has been linked to creativity and problem-solving abilities. Have you ever woken up with a solution to a problem that stumped you the previous day? That's your brain at work during REM sleep, synthesizing new ideas and solutions. Many artists, scientists, and innovators attest to the inspiration they receive during their dream cycles. The brain's ability to connect disparate ideas and form new concepts is often supercharged during this phase of sleep.

Health-conscious individuals often focus on diet and exercise, but REM sleep is an often-overlooked pillar of overall well-being. For those struggling with sleep issues, including insomniacs, it's essential to focus not just on the quantity but also the quality of sleep. Strategies like maintaining a regular sleep schedule, optimizing the sleep environment, and reducing stress can all facilitate more robust REM cycles. In turn, these lead to significant improvements in both physical and mental health.

Neglecting REM sleep can have severe repercussions. Chronic deprivation of this crucial sleep stage has been linked to impaired cognitive function, mood disorders, and even an increased risk for certain diseases. Over time, the lack of sufficient REM sleep can lead to symptoms akin to those of severe insomnia or chronic stress—including memory lapses, emotional instability, and difficulty concentrating.

In sum, REM sleep is a cornerstone of not just sleep itself but of overall health. It encompasses essential functions from cognitive processing to emotional resilience and creative problem-solving. Understanding and prioritizing REM sleep can make a significant difference in your quality of life, making you not just a better sleeper but a more focused, balanced, and creative individual. So, whether you're a health-conscious individual or an insomniac striving for better sleep, never underestimate the power of this often-mystifying aspect of the sleep cycle.

Chapter 3:
Psychological Aspects of Sleep

Understanding the psychological aspects of sleep opens the door to comprehending how our mental state interfaces with rest. This chapter delves into the intimate connection between sleep and emotions, emphasizing how stress can act as both a barrier to and a disruptor of quality shut-eye. The interplay between sleep disorders and mental health further underscores the need for psychological well-being in achieving restful slumber. Additionally, exploring the realm of dreams offers insights into our subconscious thoughts. By grasping these psychological components, we gain a holistic view of sleep, providing us with the tools to enhance our nightly rest and overall well-being.

The Connection Between Sleep and Emotions

It's no secret that the quality of our sleep greatly influences our emotional state. A single night of poor sleep can turn a normally mild-mannered person into an irritable one. But why is this the case? Emotions are intimately tied to the way our brains function, and sleep is a fundamental process that helps regulate this intricate system.

When we sleep, particularly during rapid eye movement (REM) stages, our brains are actively processing the emotional memories of the day. REM sleep helps us interpret emotional experiences and place them within our broader understanding of the world. On the flip side, the lack of sufficient REM sleep can leave us less equipped to deal with stress and more susceptible to mood swings.

Furthermore, chronic sleep deprivation has been linked to more severe emotional dysregulation. Studies have shown that insufficient sleep can exacerbate symptoms of anxiety and depression. When we don't get enough sleep, the amygdala, which controls feelings of fear and anxiety, becomes overactive. This leaves us more prone to overreacting to situations that might have otherwise seemed manageable. At the same time, the prefrontal cortex, the part of the brain that regulates emotional responses, becomes less effective.

Another key aspect is that individuals experiencing emotional distress often face difficulties sleeping, creating a vicious cycle. Stress and anxiety can lead to a hyper-aroused state that makes falling asleep challenging, which then deteriorates emotional well-being further.

It's also crucial to acknowledge the varying impacts based on individual differences. Some people are more resilient, while others might need more conscious effort to manage their emotional responses. Techniques like mindfulness and cognitive behavioral therapy can be incredibly effective in breaking the cycle of poor sleep and emotional turmoil.

Understanding the intricate bond between sleep and emotions is a step toward better emotional health and overall well-being. Prioritizing good sleep hygiene and seeking support when needed can make a substantial difference in how we navigate our emotional landscapes.

Impact of Stress on Sleep is profound and multifaceted. Stress, whether acute or chronic, greatly affects the quality and quantity of sleep, leading to disruptions that can make it difficult to achieve restorative rest. When an individual encounters stress, the body responds by activating the hypothalamic-pituitary-adrenal (HPA) axis. This response releases cortisol and other stress hormones, which prepare the body for a 'fight or flight' reaction. High cortisol levels, particularly close to bedtime, can hinder the ability to fall asleep and reduce overall sleep quality.

In the evening, our bodies naturally prepare for sleep by gradually decreasing cortisol levels, promoting relaxation and readiness for rest. However, stress can hijack this natural process, keeping cortisol levels elevated and preventing the body from transitioning into a restful state. This physiological response triggers a cycle of sleeplessness and stress, as anxiety about not being able to sleep exacerbates the stress response, creating a vicious loop.

The connection between stress and sleep occurs in both directions: not only does stress disrupt sleep, but poor sleep can also increase stress sensitivity. Lack of sufficient, quality sleep makes it harder for the body to regulate cortisol levels effectively, leaving individuals more vulnerable to stressors. Moreover, sleep deprivation impacts cognitive functions, such as memory and attention, making stressors seem more overwhelming and less manageable.

Furthermore, stress can exacerbate specific sleep disorders. For example, individuals already struggling with insomnia might find their symptoms worsening during stressful periods. Stress can also trigger episodes of sleep bruxism (teeth grinding) or exacerbate conditions like sleep apnea by increasing muscle tension.

In terms of sleep architecture, stress significantly impacts REM (Rapid Eye Movement) sleep, the stage of sleep associated with dreaming and emotional regulation. Elevated stress levels can lead to a decrease in REM sleep, impairing the brain's ability to process and manage emotions effectively. This REM sleep deficit can contribute to increased emotional reactivity during waking hours, further perpetuating the cycle of stress and poor sleep.

Physiologically, the autonomic nervous system also plays a role. Stress activates the sympathetic nervous system, responsible for the body's 'fight or flight' responses. This activation leads to increased heart rate, blood pressure, and overall arousal, making it challenging to achieve the state of calm necessary for sleep. Conversely, the parasympathetic nervous system, which promotes relaxation and rest, remains under-activated during times of stress.

Mental health conditions such as anxiety and depression, which often intertwine with stress, can further disrupt sleep patterns. Anxiety heightens the body's alertness, making it difficult to fall and stay asleep. Depression, while often associated with excessive sleep, can also cause fragmented sleep and early morning awakenings, reducing overall sleep quality.

Addressing the impact of stress on sleep requires a multifaceted approach that targets both the mind and the body. Techniques such as mindfulness meditation, deep breathing exercises, and cognitive behavioral therapy for insomnia (CBT-I) have proven effective in managing stress and improving sleep quality. Additionally, establishing a consistent sleep routine and creating a calming bedtime environment can help mitigate the effects of stress on sleep.

One powerful tool for managing stress is practicing mindfulness. Mindfulness techniques involve focusing on the present moment and accepting it without judgment. This practice helps break the cycle of worrisome thoughts that can disrupt sleep. Studies have shown that regular mindfulness practice can lower cortisol levels, reduce anxiety, and improve sleep quality by promoting relaxation and mental clarity.

Deep breathing exercises also offer significant benefits. By consciously slowing down the breath and focusing on each inhale and exhale, you can activate the parasympathetic nervous system, which promotes relaxation. This practice can lower heart rate and blood pressure, facilitating a smoother transition into sleep.

Additionally, cognitive behavioral therapy for insomnia (CBT-I) combines cognitive and behavioral strategies to address the thoughts and behaviors disrupting sleep. CBT-I is highly effective in treating insomnia and reducing the impact of stress on sleep. The cognitive component addresses the negative thought patterns that contribute to stress and sleeplessness, while the behavioral component focuses on establishing healthy sleep habits.

Creating a sleep-friendly environment is another crucial step. This includes ensuring the bedroom is dark, quiet, and cool, as well as removing electronic devices that can emit blue light and disrupt the body's natural sleep-wake cycle. Establishing a consistent bedtime routine signals to your body that it's time to wind down, making it easier to transition into sleep.

Beyond these practices, understanding and acknowledging the impact of stress on sleep is vital. By recognizing the bidirectional relationship between stress and sleep, individuals can take proactive steps to manage stress during the day and employ relaxation techniques in the evening, promoting better sleep and overall well-being.

Sleep Disorders and Mental Health are intimately connected, with each having a profound impact on the other. Stress, anxiety, and depression are some of the most recognized mental health conditions that interplay with sleep disorders. While it may seem simple at first glance, the intricacies of this relationship reveal how pivotal sleep is to our psychological well-being.

One of the most common sleep disorders is insomnia, often co-occurring with anxiety and depression. People suffering from anxiety may find that their mind races at night, keeping them awake. It's like being caught in a mental hamster wheel, where thoughts can't quiet down. Similarly, those experiencing depression frequently encounter early morning awakenings or the opposite—hypersomnia, where they sleep excessively but still feel unrefreshed. These disruptions in sleep patterns can worsen the emotional states, creating a vicious cycle.

In addition, sleep disorders such as sleep apnea can significantly impact mental health. Sleep apnea, characterized by interrupted breathing during sleep, often leads to fragmented sleep and low blood oxygen levels. This condition can result in cognitive impairments, mood disturbances, and even contribute to the development of depression. Interrupted sleep from sleep apnea doesn't allow the brain to go through the normal sleep architecture, hindering emotional regulation and cognitive function.

Considering the bidirectional relationship between sleep and mental health, it's essential to address one to improve the other. For example, treating sleep disorders can sometimes alleviate symptoms of mental health conditions. Cognitive Behavioral Therapy for Insomnia (CBT-I), a therapeutic approach discussed in Chapter 10, is effective in breaking the cycle of poor sleep and mental health issues. It's a structured program that helps individuals establish healthy sleep habits and challenges the thoughts and behaviors that prevent them from sleeping well.

Moreover, disturbed sleep patterns are not just symptoms but can also be early indicators of mental health issues. Monitoring sleep can provide crucial insights into an individual's mental state. For instance, consistent trouble falling asleep, staying asleep, or waking up too early for more than a few weeks might be a warning signal for developing anxiety or depression. Health professionals often use these patterns as part of their diagnostic criteria.

Emotional regulation is another aspect where sleep and mental health intersect. The brain's ability to regulate emotions is heavily dependent on sufficient and quality sleep, particularly REM sleep. REM sleep plays a vital role in processing emotions, and inadequate REM sleep can lead to emotional instability, irritability, and increased sensitivity to stress. This is why sleep-deprived individuals often feel emotionally on edge and are unable to cope with stress as efficiently.

It's also important to recognize that certain personality traits and mental health conditions can predispose individuals to sleep disorders. People with conditions such as bipolar disorder, schizophrenia, and ADHD often have altered sleep patterns. The manic phase of bipolar disorder, for example, might lead to reduced need for sleep, while the depressive phase might result in hypersomnia. Schizophrenics often experience fragmented sleep due to the disorder's impact on the brain's structure and function.

The impact of chronic sleep deprivation on mental health goes beyond mood disorders. Long-term sleep deprivation has been linked to cognitive decline and even neurodegenerative diseases like Alzheimer's. The brain requires sleep to clear out metabolic waste products that accumulate during the day. Without enough sleep, these toxins can build up, potentially contributing to cognitive impairment and mental decline.

Given this intricate relationship, addressing sleep disorders becomes crucial for improving mental health. Behavioral interventions, sleep hygiene practices, and medical treatments for sleep disorders can offer significant improvements. Simple steps like maintaining a regular sleep schedule, creating a restful sleep environment, and avoiding stimulants can make a big difference. Combining these with mental health support can help break the cycle of poor sleep and psychological distress.

In conclusion, the relationship between **Sleep Disorders and Mental Health** is deeply intertwined. Understanding and addressing this connection offers a pathway to better sleep quality and overall well-being. Breaking the cycle of poor sleep and mental health issues requires a comprehensive approach, but the benefits extend well beyond better sleep. They lead to improved emotional regulation, cognitive function, and a more balanced life.

Dreams and Their Meanings often captivate our imagination and curiosity. But what if they are more than just ephemeral mental images? Understanding dreams can be a valuable part of improving sleep quality and, in turn, overall well-being. By deciphering the puzzles presented in our dreams, we can unlock hidden truths about our waking lives and emotions.

Psychologists argue that dreams serve as a window into our subconscious mind. These nocturnal narratives can reflect our deepest fears, unresolved conflicts, and unspoken desires. Some theories even suggest that dreams are a mechanism through which our brain processes and organizes information accumulated during the day. When we dream, it's as if our minds are engaging in a nocturnal form of therapy, working through issues that may be difficult to address consciously.

Many people experience recurring dreams or themes. These aren't random but are often tied to ongoing stress or concerns. For instance, common themes like falling, being chased, or losing teeth might indicate feelings of powerlessness or anxiety. By recognizing these patterns, individuals can take proactive steps towards resolving underlying issues, thereby improving both mental health and sleep quality.

Symbolism in Dreams

Symbolism in dreams can convey deep and significant meanings. For instance, water often symbolizes emotions. Calm water might reflect peace, while turbulent water could signify emotional turmoil. Similarly, flying may symbolize freedom or a desire to overcome obstacles.

It's essential not to take dream symbols literally but to view them through the lens of personal experience. The context of the dream, as well as the individual's background and current life circumstances, play a crucial role in interpreting its meaning. Keeping a dream journal can be particularly helpful in identifying these symbols and understanding their relevance to one's life.

Lucid Dreaming

Lucid dreaming, the phenomenon where individuals are aware they are dreaming and can even control the narrative, offers an intriguing aspect of dream analysis. There is evidence suggesting that lucid dreaming can be a powerful tool for personal growth and problem-solving.

For those suffering from nightmares, particularly those related to trauma or anxiety, learning to induce lucid dreams can be therapeutic. This technique allows individuals to confront and alter distressing dream scenarios, reducing their nighttime awakenings and improving sleep quality.

Scientific Perspectives

From a scientific perspective, dreams are a subject of intense study. The brain's activity during REM (Rapid Eye Movement) sleep is notably different from other sleep stages and is thought to be when most dreaming occurs. Neuroimaging studies have shown that areas of the brain associated with emotions and memory, such as the amygdala and hippocampus, are particularly active during REM sleep.

Researchers suggest that this high level of activity indicates a role for dreams in emotional regulation and memory consolidation. By processing memories and emotions in the relatively safe environment of a dream, the brain might be better equipped to handle these elements in waking life.

Cultural Interpretations

Cultural differences also play a significant role in how dreams are perceived and interpreted. In some cultures, dreams are seen as prophetic, offering glimpses into the future or messages from the divine. In others, they're considered purely a psychological phenomenon, reflecting the dreamer's mind.

Regardless of the cultural lens, what remains consistent is the understanding that dreams have an impact. By paying attention to these nocturnal visions, individuals may gain valuable insights into their emotional and mental states.

Practical Applications

Practical applications of dream analysis include improving personal relationships and making crucial life decisions. For example, a recurring dream about conflict with a loved one may signal unresolved disputes that need addressing. Similarly, dreams that evoke strong emotions can guide individuals toward understanding what truly matters to them.

Moreover, dream analysis can be incorporated into therapeutic practices. Therapists may use a client's dreams to gain insights into their subconscious mind, helping to illuminate issues that might be affecting their mental health and sleep.

Dream Journals

One of the most effective ways to understand and interpret dreams is to maintain a dream journal. By recording dreams immediately upon waking, individuals can capture details that might otherwise be forgotten. Over time, patterns and common themes may emerge, providing a clearer picture of the dreamer's subconscious mind.

This practice can also help in identifying specific triggers or stressors that need to be addressed. For example, a series of anxiety-filled dreams might correlate with a stressful period at work or personal life challenges, prompting focused interventions to alleviate stress.

Final Thoughts

In a broader sense, understanding dreams and their meanings can be akin to embarking on a journey of self-discovery. It's an ongoing process that requires patience and self-reflection. The insights gained from this exploration can lead to improved emotional health, better relationships, and enhanced sleep quality.

Ultimately, dreams hold powerful keys to understanding our inner selves. By paying attention to these nightly narratives and seeking to interpret their meanings, we can gain a deeper comprehension of our emotions and impulses. This continued effort not only enriches our waking lives but also paves the way for restful and restorative sleep.

Chapter 4:
The Health Benefits of Good Sleep

Good sleep is like a powerful elixir for your body and mind, offering an array of health benefits that can't be ignored. Physically, sufficient sleep enhances immune function, helping your body fend off illnesses in ways you're often not even aware of. Quality sleep is also essential for maintaining a healthy weight, as it regulates hormones that control hunger and satiety, thereby playing a critical role in weight management. Additionally, well-rested individuals have better heart health, as sleep supports cardiovascular function by maintaining blood pressure levels and reducing stress on the heart. On the cognitive side, good sleep sharpens memory, enhances problem-solving abilities, and boosts overall mental clarity. Thus, prioritizing sleep is not just about feeling rested; it's a cornerstone for long-term health and well-being.

Physical Health and Immune Function

You might not immediately connect a good night's sleep with your immune system, but the two are deeply intertwined. When you sleep, your body goes into a kind of maintenance mode, repairing cells, restoring energy, and releasing crucial hormones. This restorative process is essential for maintaining your overall physical health. For example, during deep sleep stages, the body produces cytokines, a type of protein that fights infections and inflammation. This is when your immune system gets its crucial "service and repair."

Inadequate sleep can weaken this defense mechanism. People who don't get enough sleep are more likely to fall ill after being exposed to a virus, such as the common cold. Conversely, adequate sleep boosts your immune system's memory. This means that it can "remember" pathogens it has encountered before and fight them off more effectively.

But it doesn't end there. Good sleep has a myriad of benefits that span various aspects of physical health. For instance, it helps regulate your metabolism and can even affect your weight. When you're sleep-deprived, your body's ability to metabolize glucose diminishes, leading to an increased risk of conditions like type 2 diabetes. You're also more likely to crave high-carb and sugary foods when you're tired, further exacerbating weight issues.

Moreover, quality sleep has been linked to a healthier heart. Poor sleep can lead to increased blood pressure and higher levels of inflammation, both of which are risk factors for heart disease. During sleep, your heart rate and blood pressure naturally decrease, giving your cardiovascular system a much-needed rest. This nightly respite helps to maintain heart health over the long term.

So, by prioritizing good sleep, you're not just feeling more refreshed and alert. You're actively contributing to your long-term physical health and boosting your immune system's ability to ward off illness. Quite literally, sleep could be the best medicine for a healthier, more resilient you.

Sleep and Weight Management is a fascinating nexus of health science and behavioral insights. As researchers have delved deeper into the human body's interconnected systems, it has become clear that the relationship between sleep and weight management is not just about the number of hours you spend in bed. Understanding this connection could be the key to not only better sleep but also to maintaining a healthy weight.

Why Sleep Matters for Weight Management

Sleep is more than an opportunity for your body to rest—it's a dynamic process that affects almost every system. One of the most compelling ways sleep influences weight is through its impact on hunger hormones, namely ghrelin and leptin. Ghrelin, known as the "hunger hormone," increases your appetite, while leptin tells your brain when you're full. Sleep deprivation tips this hormonal balance, raising ghrelin levels and reducing leptin levels, making it much easier to overeat.

Moreover, sleep deprivation affects insulin sensitivity. Insulin is crucial for regulating your blood sugar levels. When you're sleep-deprived, your body becomes less effective at using insulin, leading to higher blood sugar levels and an increased risk of type 2 diabetes. This insulin resistance also promotes fat storage, particularly around the abdomen, exacerbating weight gain.

The Role of Sleep Duration and Quality

Both sleep duration and quality are pivotal in weight management. The National Sleep Foundation recommends that adults aim for 7-9 hours of sleep per night. But it's not just about quantity; quality counts just as much. Fragmented sleep, often caused by sleep disorders like sleep apnea, can be just as detrimental as insufficient sleep. Poor sleep quality disrupts the balance of hunger hormones and reduces insulin sensitivity, much like sleep deprivation does.

Interestingly, research has shown a linear relationship between sleep and weight. For each hour of sleep lost, the risk of obesity increases. This association is particularly pronounced in children and adolescents, making good sleep hygiene critically important during growth years.

Psychological Aspects and Eating Behaviors

Sleep loss doesn't just change how your body responds to food; it also affects how you think about and choose food. When you're tired, your brain's reward centers are more likely to crave high-calorie, high-fat foods. This isn't just a quirk—it's a survival mechanism. Your body craves quick energy to combat fatigue, leading you to reach for sugary snacks and fast food more often.

The implications are significant: tiredness undermines willpower. Studies have shown that people are less likely to cook healthy meals and more likely to engage in emotional eating when they haven't had adequate sleep. This vicious cycle means that poor sleep leads to poor eating habits, which in turn can lead to weight gain and further sleep issues.

Physical Health Connections

Weight gain and obesity are associated with an increased risk of various health conditions, from cardiovascular diseases to metabolic disorders. But poor sleep exacerbates these risks. For instance, sleep apnea, a condition often linked to obesity, can create a feedback loop where poor sleep quality perpetuates weight gain, which in turn worsens sleep apnea.

Additionally, inflammation, a common factor in chronic conditions such as diabetes and heart disease, can be exacerbated by poor sleep. Inflammatory markers like C-reactive protein are elevated in individuals who are sleep-deprived, which can further complicate weight management efforts and overall health.

Practical Steps for Better Sleep and Weight Management

So, how can you break the cycle and foster a healthier relationship between sleep and weight? Here are a few actionable tips:

1. **Create a Sleep Schedule:** Go to bed and wake up at the same time every day, even on weekends. This helps regulate your body's internal clock and improves the quality of your sleep.

2. **Prioritize Sleep Hygiene:** Your sleep environment is crucial. Ensure your bedroom is cool, dark, and quiet. Invest in a comfortable mattress and pillows.

3. **Mind Your Diet:** Pay attention to what and when you eat. Avoid large meals, caffeine, and alcohol right before bedtime as they can disrupt your sleep.

4. **Exercise Regularly:** Physical activity can help you fall asleep faster and enjoy deeper sleep. However, avoid vigorous exercise close to bedtime.

5. **Manage Stress:** Chronic stress can lead to sleep problems. Techniques such as meditation, yoga, and deep breathing can help manage stress and improve sleep quality.

Consult Professionals

If you find that despite making lifestyle changes, you are still struggling with both sleep and weight management, it might be time to seek professional advice. Sleep studies can uncover disorders like sleep apnea, and a nutritionist can provide tailored advice to help you manage your weight without compromising your sleep quality.

Ultimately, the interplay between sleep and weight management is complex but not insurmountable. By understanding and addressing the factors that link these two crucial aspects of health, you can take significant strides towards achieving both better sleep and a healthier weight, thereby improving your overall well-being.

Sleep's Role in Heart Health cannot be overstated. It serves as a foundation for numerous physiological functions, including cardiovascular well-being. The body isn't just resting during sleep; it's undergoing vital processes crucial for maintaining heart health. It's important to understand that moments of sleep translate into a robust heart, mainly due to the interplay of sleep stages and their impact on bodily functions.

When we enter deep sleep, the heart rate and blood pressure drop significantly compared to waking hours. This reduction gives the heart a much-needed rest. During the non-rapid eye movement (NREM) stages, particularly stages 3 and 4, the body enters what some call "deep sleep" or "slow-wave sleep." It's in these stages that the sympathetic nervous system—responsible for our fight-or-flight responses—calms down. This deceleration lowers blood pressure, decreases stress hormone levels, and promotes relaxation and restoration in vascular tissues.

The role of *REM sleep* shouldn't be dismissed either. During REM sleep, blood pressure and heart rate can increase, but this fluctuation is quite different from the stress-induced spikes we experience during the day. Think of it as a form of cardiovascular exercise; the heart gets a gentle workout that promotes flexibility and adaptability. This adaptive function may play a role in regulating the cardiovascular system and keeping it responsive to daily stresses.

It's worth noting that sleep's impact on heart health is bidirectional. Poor sleep can exacerbate heart-related issues, while heart problems can, in turn, interfere with one's sleep. Chronic sleep deprivation has been linked with a plethora of heart-related conditions, including hypertension, coronary artery disease, and heart failure. Several studies suggest that people who don't get enough sleep are at a higher risk of developing these conditions, primarily due to increased levels of inflammation, stress hormones, and disrupted metabolic functions.

One significant mechanism by which poor sleep affects heart health is through the regulation of insulin and glucose. Sleep deprivation impairs glucose metabolism and insulin sensitivity, leading to higher blood sugar levels. Over time, this can result in type 2 diabetes, another risk factor for cardiovascular disease. Elevated blood sugar levels contribute to chronic inflammation, which in turn damages the arteries and facilitates the development of plaques, leading to atherosclerosis.

Consider also the role of hormonal balance in sleep and heart health. The body's production of cortisol, a hormone released in response to stress, follows a circadian rhythm. Typically, cortisol levels peak in the morning upon waking and decline throughout the day. Poor sleep patterns can disrupt this natural cycle, causing cortisol to remain elevated, which poses risks such as high blood pressure and increased arterial stiffness. High cortisol levels are particularly detrimental because they provoke the body to retain sodium, thereby increasing blood volume and forcing the heart to work harder.

Obstructive sleep apnea (OSA) is another critical factor linking sleep and heart health. In OSA, interrupted breathing during sleep leads to repeated episodes of hypoxia, putting immense strain on the cardiovascular system. Each cessation of breath causes oxygen levels to drop and carbon dioxide levels to rise, stimulating an alert response. This arousal causes sudden spikes in heart rate and blood pressure. Over time, these fluctuations can contribute to hypertension, arrhythmias, and even increase the risk of stroke and heart attack. Effective treatment of OSA has been shown to significantly improve heart health, underscoring the importance of addressing sleep disorders.

Managing sleep to enhance heart health involves multiple strategies, beginning with adhering to good sleep hygiene. This isn't just about avoiding caffeine or creating a dark sleep environment; it's about consistency, reducing stress, and creating rituals that signal the body it's time to wind down. Techniques such as mindfulness meditation and controlled breathing exercises can be particularly effective in reducing pre-sleep anxiety, which often disrupts heart-friendly sleep stages.

Equally important are regular physical activity and a balanced diet, which contribute significantly to improved sleep and, by extension, better heart health. Engaging in moderate exercise can lower blood pressure and improve sleep quality, while a diet rich in omega-3 fatty acids, fiber, and antioxidants supports overall cardiovascular function and mitigates the risk of developing heart-related issues linked with poor sleep.

It's essential to recognize the symbiotic relationship between sleep and heart health as a cornerstone for overall well-being. When prioritizing sleep, you're not just enhancing your cognitive functions and emotional balance, you're also fortifying your heart. By adopting and adhering to practices that promote quality sleep, you're investing in one of the most vital aspects of health there is—the longevity and functionality of your heart.

Cognitive Benefits

When it comes to cognitive function, good sleep is nothing short of a game-changer. The brain is like a complex orchestra, and sleep is its quintessential conductor, ensuring that every section comes in at the right moment. Proper sleep can significantly enhance your ability to concentrate and focus, making everyday tasks seem more manageable and less daunting. After a well-rested night, you might notice that your mind feels sharper, thoughts flow more freely, and even remembering that long list of errands becomes a cinch.

One of the striking cognitive benefits of good sleep is improved memory. During various sleep stages, particularly REM sleep, the brain goes through a process called memory consolidation. This is when your brain takes what you've learned throughout the day and solidifies it, transferring short-term memory to long-term storage. Whether you're learning a new language, picking up a musical instrument, or simply aiming to remember names and faces better, sleep is an essential part of the equation that ensures these new experiences and informations are stored efficiently.

Moreover, creativity thrives on a good night's sleep. This happens because sleep enables the brain to make new and unexpected connections between ideas, which is the cornerstone of creative thinking. You might have experienced going to bed with a problem and waking up with an "aha" moment or a fresh perspective.

Decision-making abilities also benefit enormously from proper rest. When you're sleep-deprived, the prefrontal cortex, responsible for logical thinking and decision-making, operates less effectively. This impaired function can lead to poor choices, whether in professional settings or personal life. On the other hand, adequate sleep empowers you to evaluate situations more clearly and make decisions that are well-informed and balanced.

The interplay between sleep and emotional intelligence is another facet not to be overlooked. Good sleep allows the brain to regulate emotions better, enhancing empathy and reducing the likelihood of emotional outbursts or mood swings. This becomes particularly important in maintaining healthy interpersonal relationships, whether at work, home, or social settings. In essence, getting quality sleep doesn't just rest your body, it rejuvenates your mind, bolstering every cognitive function you rely on daily.

Chapter 5:
Common Sleep Disorders

Sleep disorders can profoundly impact your quality of life, turning rest into a nightly battle rather than a period of rejuvenation. Insomnia, one of the most prevalent sleep disorders, manifests in various forms from difficulty falling asleep to staying asleep, often exacerbated by factors such as stress or underlying health conditions. Sleep apnea, characterized by interrupted breathing during sleep, can leave you feeling exhausted despite a seemingly full night's rest, and it's often marked by loud snoring or abrupt awakenings. Restless Leg Syndrome introduces an uncontrollable urge to move your legs, making it challenging to drift into sleep. Understanding these common sleep disorders empowers you to seek suitable interventions, whether it involves lifestyle changes, medical treatment, or cognitive techniques, ultimately aiming to reclaim a restful night's sleep.

Insomnia

Insomnia is one of the most prevalent sleep disorders affecting millions of people around the globe. It manifests as difficulty falling asleep, staying asleep, or both. While it's completely normal to have occasional sleepless nights, chronic insomnia is a more severe condition that can significantly impact one's overall quality of life and well-being.

Understanding insomnia requires delving into its complexities, which include its various types, potential causes, and associated risk factors. There are two primary types of insomnia: acute and chronic.

Acute insomnia is short-term and often triggered by specific events or stressors, like an important meeting the next day or a traumatic event. Chronic insomnia, on the other hand, lasts for more than three months and can result from a variety of underlying issues such as medical conditions, mental health disorders, or poor sleep habits.

The causes of insomnia are diverse. Psychological factors like anxiety, stress, and depression are common contributors. Additionally, lifestyle choices such as irregular sleep schedules, excessive screen time, and consumption of caffeine or alcohol can exacerbate the problem. Physical conditions, such as chronic pain or respiratory issues, further complicate one's ability to achieve restful sleep.

It's also important to consider the risk factors associated with insomnia. Certain demographics are more susceptible, like older adults, due both to changes in sleep patterns and the prevalence of chronic health conditions that can disrupt sleep. Women are also at higher risk, particularly during hormonal transitions like menopause. Understanding these risk factors can aid in the early identification and management of the disorder.

Living with insomnia can be frustrating and exhausting, but it's crucial to remember that effective treatments are available. Cognitive Behavioral Therapy for Insomnia (CBT-I) is a highly recommended method, focusing on changing sleep habits and misconceptions that fuel insomnia. Medication is another option, though it should be used sparingly and under the guidance of a healthcare provider.

In summary, insomnia is a multifaceted condition requiring a holistic approach for improvement. By addressing both the psychological and physiological aspects, individuals can reclaim their nights and enjoy the numerous health benefits of a good night's sleep.

Types of Insomnia

Insomnia isn't just a matter of lying awake at night, staring at the ceiling. The condition manifests itself in various ways, and identifying the type you're experiencing is key to finding the right solution. Broadly speaking, insomnia can be categorized into three main types: transient, acute, and chronic. Understanding these distinctions can guide both diagnosis and treatment, helping individuals tailor strategies to improve their sleep quality and overall well-being.

First, let's delve into transient insomnia. This type is often linked to temporary changes in a person's lifestyle or environment. For instance, travel-related disruptions like jet lag, stress about an upcoming presentation, or even a noisy neighborhood can trigger transient insomnia. The good news? It's usually short-lived, lasting less than a week. In many cases, once the triggering factor is resolved, normal sleep patterns return. However, if ignored, even transient insomnia can set the stage for more severe sleep problems down the line.

Acute insomnia, on the other hand, might last for several weeks and often arises due to significant life events such as the death of a loved one, a job loss, or a stressful project at work. This type of insomnia is more severe than transient insomnia but still considered short-term. Individuals may find themselves lying awake at night, ruminating over the stressful event, and unable to quiet their minds. Acute insomnia can disrupt daily functioning and weaken the immune system, making it crucial to address the root causes promptly.

Chronic insomnia is where things get more complicated. *This* form of insomnia persists at least three nights a week over a period of three months or longer. Chronic insomnia is often multifaceted, involving a blend of psychological, physiological, and behavioral factors. While stress and anxiety are common triggers, chronic insomnia can also be linked to underlying health problems like sleep apnea or restless leg syndrome. In some cases, poor sleep habits themselves perpetuate the cycle, with individuals developing anxiety about not being able to sleep, thereby worsening their insomnia.

Each type of insomnia has unique characteristics and implications. *It's essential to recognize these nuances* to devise effective interventions. While transient insomnia might be managed with temporary adjustments in sleep environment or routines, acute and chronic insomnia often require more comprehensive approaches, including cognitive-behavioral strategies, medication, or lifestyle changes.

No conversation about insomnia would be complete without acknowledging secondary insomnia, a condition driven by other health issues. Whether it's due to chronic pain, respiratory problems, or hormonal changes, secondary insomnia stems from another medical condition that directly or indirectly disrupts sleep. Addressing the primary condition often helps alleviate the insomnia, though sometimes targeted sleep interventions might still be necessary.

Even among these primary categories, insomnia can further subdivide based on its specific manifestations, such as sleep-onset insomnia and sleep-maintenance insomnia. Sleep-onset insomnia involves difficulty falling asleep at the beginning of the night and is often associated with anxiety disorders or poor sleep hygiene, such as excessive use of screens before bed. Sleep-maintenance insomnia, conversely, is the difficulty in staying asleep. People with this condition may fall asleep easily but wake up frequently during the night, often struggling to fall back asleep. This form is more commonly linked to conditions like depression and age-related changes.

Then, there's mixed insomnia—a combination of sleep-onset and sleep-maintenance issues. Such cases can be particularly challenging, as they may require a multifaceted treatment approach that addresses both falling and staying asleep. The complexity of mixed insomnia often necessitates a comprehensive evaluation by a sleep specialist to identify all contributing factors and effectively tailor interventions.

Another variant is paradoxical insomnia, where individuals perceive their sleep as insufficient despite objective evidence to the contrary. Also known as sleep state misperception, this form involves a mismatch between how long individuals think they've slept and how long they actually have. People with paradoxical insomnia often feel exhausted and irritable during the day, even though sleep studies may show normal sleep duration and quality. Cognitive-behavioral therapy (CBT) is particularly effective for this type, helping to recalibrate sleep perceptions and improve sleep-related anxiety.

It's also worth mentioning idiopathic insomnia, a rare condition that typically starts in childhood and persists throughout a person's life without any obvious external cause. The origins of idiopathic insomnia are not well understood, but it is believed to involve an inherent abnormality in the sleep-wake cycle. Treatment options for idiopathic insomnia are often limited and focus on managing symptoms to improve sleep quality and daytime functioning.

Different types of insomnia often require tailored treatments. Short-term strategies like improving sleep hygiene or using sleep aids might work for transient insomnia. In cases of acute insomnia, stress management techniques, such as mindfulness and relaxation exercises, can be beneficial. Chronic insomnia often requires a multi-pronged approach, combining cognitive-behavioral therapy (CBT), medicinal interventions, and lifestyle modifications for effective treatment.

The impact of insomnia goes beyond just feeling tired. Chronic sleep deprivation can lead to significant health issues, including cardiovascular diseases, obesity, and weakened immune function. Moreover, it can affect mental health, contributing to conditions like depression and anxiety. This makes addressing insomnia not just a necessity for better sleep, but a crucial aspect of overall health and well-being.

Seek professional help if your insomnia symptoms persist. A healthcare provider specialized in sleep medicine can perform a thorough evaluation, identify underlying causes, and recommend appropriate treatments. By understanding the type of insomnia you're dealing with, you can take targeted steps towards better sleep and a healthier life.

Causes and Risk Factors within the realm of insomnia are multifaceted and complex, encompassing both physiological and psychological elements. It's no surprise that understanding these causes can often feel like peeling an onion—revealing layers upon layers of contributing factors. Identifying these underlying issues is a crucial step toward mitigating the pervasive impact insomnia can have on one's life.

One major factor to consider is the influence of stress and anxiety. Many individuals experience difficulty falling or staying asleep when preoccupied with worry. Stress triggers the body's "fight or flight" response, releasing cortisol and other stress hormones, which can disrupt the natural sleep cycle. Prolonged stress can lead to chronic insomnia, establishing a vicious cycle where lack of sleep contributes to greater stress and further sleep issues.

Another significant physiological contributor is the imbalance of key neurotransmitters in the brain. Neurotransmitters such as serotonin and gamma-aminobutyric acid (GABA) play critical roles in promoting relaxation and sleep. An imbalance in these chemicals can lead to difficulty winding down at night and staying asleep, even if you're physically exhausted.

Medical conditions also come into play. Chronic pain, respiratory diseases such as asthma or chronic obstructive pulmonary disease

(COPD), and even gastrointestinal issues like acid reflux can interfere with sleep. These conditions cause discomfort or pain that makes it difficult to maintain a restful sleep state. Consider, too, the impact of neurological conditions like Parkinson's disease or Alzheimer's, which can disrupt the brain's regulation of sleep.

Medications prescribed for various ailments can further muddy the waters. Some drugs have side effects that include insomnia or disturbed sleep patterns. Stimulants such as those used for treating ADHD and certain antidepressants can interfere with sleep architecture, robbing individuals of restorative rest. Even over-the-counter medications like decongestants contain stimulants that can keep you up at night.

Behavioral factors can't be overlooked. Irregular sleep schedules, often seen in shift workers or those with erratic work hours, can disrupt circadian rhythms. This misalignment leads to difficulties initiating and maintaining sleep. Additionally, lifestyle choices such as excessive consumption of caffeine, alcohol, or nicotine can be detrimental. Caffeine and nicotine are stimulants, while alcohol, although it might initially induce drowsiness, disrupts night-time sleep stages and prevents deep sleep.

Environmental factors also wield significant influence. A noisy or uncomfortable sleep setting can prove to be a substantial barrier to achieving quality sleep. Light pollution, uncomfortable bedding, or room temperature that's either too hot or too cold can all disrupt sleep. Even the psychological association of the bedroom with work or stress, rather than rest and relaxation, can impact one's ability to fall asleep.

Age and hormonal changes are worth mentioning. Insomnia becomes more prevalent as we age, potentially due to changes in sleep structure and circadian rhythms. Post-menopausal women experience hormonal shifts that can cause hot flashes and night sweats, contributing to sleep disturbances. Pregnant women, too, face unique challenges such as hormonal changes and physical discomfort that affect sleep.

Sometimes, the risk factors are rooted in mental health. Conditions like depression, anxiety disorders, and bipolar disorder are commonly linked with sleep difficulties. Insomnia and mental health issues often exist in a bidirectional relationship, each exacerbating the other. For instance, lack of sleep can worsen depressive symptoms, while the symptoms of depression can make it harder to sleep.

Genetics may also play a role. Some individuals have a family history of insomnia, suggesting a predisposition to sleep difficulties. While the exact genetic mechanisms are still under study, it's clear that genetics can influence one's propensity for sleep disorders.

In summary, insomnia doesn't stem from a single cause but rather a convergence of multiple factors. Identifying and addressing these risk factors requires a comprehensive approach, understanding that each individual may experience a unique combination of triggers. Awareness of these varied influences can empower individuals to take actionable steps towards improving their sleep quality.

Sleep Apnea

Sleep apnea is one of the most common yet often underdiagnosed sleep disorders. It disrupts your sleep by causing frequent pauses in breathing, which can last from a few seconds to minutes. These interruptions can occur upwards of 30 times or more in an hour. Typically, normal breathing restarts with a loud snore or choking sound. This condition can leave you feeling tired and groggy throughout the day, making everyday tasks seem more daunting.

There are three main types of sleep apnea: obstructive sleep apnea (OSA), central sleep apnea (CSA), and complex sleep apnea syndrome. The most common form, OSA, occurs when the throat muscles intermittently relax and block the airway. In contrast, CSA involves a failure in the brain's signals to muscles that control breathing. Complex sleep apnea is a combination of both.

Notably, sleep apnea doesn't just compromise sleep quality; it impacts overall health. Untreated sleep apnea elevates the risk of high blood pressure, heart disease, stroke, diabetes, and even certain types of cancer. Furthermore, it can contribute to mental health issues like depression and anxiety, amplifying the urgency to get a proper diagnosis and treatment.

The symptoms of sleep apnea extend beyond loud snoring. You might experience episodes of gasping for breath during sleep, waking up with a dry mouth or sore throat, morning headaches, difficulty staying asleep (insomnia), and excessive daytime sleepiness (hypersomnia). If these symptoms resonate with your experiences, it may be time to consult a healthcare provider.

Diagnosis often involves a recommendation for a polysomnography or home sleep tests, which monitor your sleep patterns, blood oxygen levels, heart rate, and breathing activity. Based on the diagnosis, treatment options can vary from lifestyle changes, such as weight loss and

quitting smoking, to medical interventions like Continuous Positive Airway Pressure (CPAP) therapy, oral appliances, or even surgery.

Recognizing and addressing sleep apnea can significantly enhance the quality of your sleep and, consequently, your overall well-being. If you or someone you know wrestles with these symptoms, seeking medical advice is a vital step toward better sleep health.

Symptoms and Diagnosis of sleep apnea can be subtle or glaring, varying widely among individuals. Some people might notice frequent episodes of shallow breathing or prolonged pauses between breaths during sleep, often observable by a partner. Others might wake up abruptly, gasping for air or feeling short of breath. Daytime sleepiness is a hallmark symptom, often so severe that it hampers daily activities, making it difficult to stay awake during meetings, while reading, or even when driving.

A telltale symptom is loud and chronic snoring, typically punctuated by periods of silence followed by loud gasps or choking sounds. This disrupted sleep leads to fragmented and non-restorative rest, prompting excessive fatigue during the day. Headaches upon waking, dry mouth, and difficulty concentrating are also common. Many individuals report waking with a sensation of choking or with a sore throat, indicative of frequent awakenings during the night. Mood changes, such as irritability or depression, can also arise from the chronic fatigue brought on by this condition.

Possible diagnostic indicators extend beyond subjective symptoms. Physical manifestations such as hypertension, obesity, and large neck circumference can be significant markers that prompt further investigation. Healthcare providers often begin with a detailed medical and sleep history, questioning about sleep habits, daytime alertness, and any witnessed apneic episodes. It's crucial to understand that these symptoms may overlap with other disorders, hence a thorough evaluation is essential for accurate diagnosis.

Polysomnography, or an overnight sleep study, remains the gold standard for diagnosing sleep apnea. This non-invasive test monitors various physiological signals during sleep, including brain waves, eye movements, heart rate, blood oxygen levels, airflow, and breathing patterns. Patients are typically observed for a full night at a sleep center, though home sleep tests are becoming more common and convenient. These home tests involve wearing a simplified monitor that tracks specific physiological parameters, focusing primarily on apnea events and blood oxygenation levels.

Continuous monitoring allows for detailed analysis, revealing the frequency and severity of apneic events. Apnea-Hypopnea Index (AHI), a critical metric, quantifies the number of apnea and hypopnea episodes per hour of sleep, thus categorizing the condition's severity. Mild sleep apnea is identified with an AHI of 5-15 events per hour, moderate is 15-30, and severe is above 30. These measurements are integral in formulating an effective treatment plan.

The next diagnostic step might include referral to a specialist, such as a sleep medicine physician or an otolaryngologist. Further assessments could involve imaging studies like an MRI or CT scan to visualize the upper airway structure and identify any anatomical abnormalities contributing to obstructed breathing. Regular follow-up appointments ensure that the chosen treatment effectively mitigates symptoms and improves sleep quality.

Other diagnostic tools might include questionnaires like the Epworth Sleepiness Scale or the Berlin Questionnaire. These self-reported tools help quantify daytime sleepiness and identify those at high risk for sleep apnea. By combining subjective reports with objective findings, healthcare providers can construct a comprehensive view of the patient's sleep health.

Early diagnosis is crucial to prevent severe complications, such as cardiovascular diseases, diabetes, and cognitive impairments. Addressing the symptoms and securing a thorough diagnosis can pave the way for treatments that dramatically enhance not only sleep quality but overall well-being.

Treatment Options for sleep apnea are varied, each aimed at addressing the unique causes and manifestations of the disorder. The most widely prescribed treatment is Continuous Positive Airway Pressure (CPAP). This method involves wearing a mask over the nose (and sometimes the mouth) while sleeping. The mask is connected to a machine that delivers a continuous stream of air, keeping the airways open. For many people, CPAP can significantly reduce the number of sleep apnea events, leading to better sleep quality and improved daytime alertness.

For those who find CPAP uncomfortable or ineffective, there are alternative solutions. One such option is the use of an oral appliance. These custom-fitted devices, made by a dentist, help keep the throat open by repositioning the lower jaw and tongue. Oral appliances have been especially effective in treating mild to moderate sleep apnea and are often favored for their portability and ease of use.

Surgical options also exist for individuals who do not respond to CPAP or oral appliances. Various surgical interventions aim to remove or shrink tissues that obstruct the airways, or to reposition structures such as the jaw. Uvulopalatopharyngoplasty (UPPP), for instance, involves removing excess tissue from the throat to widen the airway. Another procedure, called genioglossus advancement, repositions the muscle that attaches the tongue to alleviate airway blockages. While surgery can be effective, it comes with risks and a longer recovery period, so it's generally considered only after other treatments have failed.

In addition to these primary treatments, lifestyle changes can play a crucial role in managing sleep apnea. Weight loss is particularly beneficial because excess weight, especially around the neck, can contribute to airway obstruction. Even a modest reduction in weight can lead to substantial improvements in symptoms. Alcohol and sedative use should also be minimized, as they relax the throat muscles, worsening sleep apnea. Furthermore, sleeping on the side rather than the back can prevent the tongue and soft tissues from collapsing into the airway.

Certain complementary therapies have also shown promise. Positional therapy, for example, involves using devices that help keep you sleeping in a side position rather than on your back. This can be as simple as a special pillow or as sophisticated as an electronic device that subtly vibrates when you move onto your back. Additionally, practices like yoga and other exercises that improve muscle tone can help maintain open airways during sleep.

For some, the inclusion of supplemental oxygen may be beneficial, particularly if they have concurrent lung disorders. Oxygen therapy ensures that enough oxygen enters the bloodstream, even if breathing is temporarily obstructed during sleep. It's typically provided through nasal prongs or a face mask, and the flow can be adjusted as needed.

Lastly, emerging technologies and treatments are continually being researched and developed. For example, hypoglossal nerve stimulation involves a small device implanted in the chest that delivers electrical impulses to the nerve controlling tongue movements, preventing airway collapse. This method shows promise for individuals who have not found success with traditional treatments.

In conclusion, there isn't a one-size-fits-all remedy for sleep apnea. Effective management often requires a combination of treatments tailored to the individual's specific condition and lifestyle. Consulting with a healthcare provider who specializes in sleep disorders is essential for determining the most suitable approach.

Restless Leg Syndrome

Restless Leg Syndrome (RLS), also known as Willis-Ekbom Disease, is a neurological disorder that's often categorized by an irresistible urge to move the legs. This urge is frequently accompanied by unpleasant sensations such as tingling, creeping, or crawling feelings. These sensations usually occur in the late afternoon or evening hours and are most severe at night when a person is resting, such as sitting or lying in bed. Because of this, RLS can significantly impact sleep quality, making it one of the most challenging sleep disorders to manage.

The exact cause of RLS remains unknown, but research suggests that genetics may play a role, as it often runs in families. Additionally, abnormalities in the brain's dopamine pathways, which are crucial for controlling movement, may contribute to the condition. Iron deficiency is another often-cited factor, possibly due to its role in dopamine production and function. Some studies also indicate that RLS may be linked to pregnancy, particularly in the last trimester, and to certain chronic diseases, such as kidney failure, diabetes, and peripheral neuropathy.

Living with RLS can be distressing. The symptoms not only interfere with falling asleep but can also cause frequent awakenings during the night. For many, this leads to chronic sleep deprivation, which further exacerbates the issue. Sleep deprivation, in turn, can affect various aspects of daily life, reducing cognitive function, impairing work performance, and increasing the risk of anxiety and depression. Because of the cyclical nature of this issue, managing RLS is critical for improving overall well-being and sleep quality.

Diagnosing Restless Leg Syndrome primarily involves a clinical evaluation. Healthcare providers often look for the four essential features: an overwhelming urge to move the legs, symptoms that begin or worsen during periods of rest or inactivity, partial or temporary relief by movement, and worsening symptoms in the evening or night. These criteria help distinguish RLS from other conditions that may cause similar symptoms, such as leg cramps or neuropathy.

There's no one-size-fits-all treatment for RLS, as the severity and underlying causes can vary. Non-pharmacological approaches often serve as the first line of defense. These can include lifestyle modifications such as establishing a regular sleep routine, keeping the legs active during the day, and practicing good sleep hygiene. Additionally, some individuals find that moderate exercise, stretching, and activities like yoga help alleviate symptoms. It's essential to avoid potential triggers, such as caffeine, alcohol, and nicotine, as these can worsen RLS symptoms.

When lifestyle changes aren't sufficient, medications might be prescribed. Dopaminergic agents, which increase dopamine levels in the brain, are commonly used and can be highly effective for many patients. Other options include anti-seizure medications, which may reduce sensory disturbances and muscle relaxants or sleep medications for those with severe insomnia. However, it's crucial to consult with a healthcare provider to tailor the treatment to individual needs and monitor potential side effects.

Iron supplementation is another potential treatment, particularly for those with iron deficiency. Before starting any supplementation, it's important to conduct blood tests to confirm low iron levels. Some individuals may also explore alternative therapies, such as acupuncture or massage, although more research is needed to validate their efficacy specifically for RLS.

Coping with RLS requires understanding and support. The unpredictability of symptoms can be frustrating, not just for those who suffer from the condition but also for their loved ones. Open communication about the limitations and challenges posed by RLS can help alleviate some of the emotional stress. Support groups, both in-person and online, can offer valuable advice and encouragement to those affected.

In conclusion, Restless Leg Syndrome is a multifaceted disorder with significant implications for sleep quality and overall health. By recognizing the symptoms, seeking appropriate medical advice, and making informed lifestyle choices, individuals can better manage their condition. As research continues to evolve, more effective treatments and a deeper understanding of RLS will hopefully emerge, bringing much-needed relief to those who suffer from this challenging disorder.

Chapter 6:
Diagnosing Sleep Issues

As you journey deeper into understanding sleep, diagnosing sleep issues becomes crucial. It's not just about tallying the hours you spend in bed; it's about analyzing the quality of those hours. Here, various diagnostic tools come into play, from polysomnography conducted in sleep labs to home sleep tests you can administer yourself. Identifying indicators such as breathing irregularities, leg movements, and brainwave patterns can help pinpoint disturbances in your sleep architecture. Recognizing when to consult a specialist becomes the next logical step, especially if symptoms persist despite improved sleep hygiene. These methods collectively provide a comprehensive picture, helping you and your healthcare provider devise effective strategies to enhance your sleep quality and, ultimately, your overall well-being.

Sleep Studies and Tests

When it comes to diagnosing sleep issues, sleep studies and tests play a vital role. One of the most common and comprehensive tests is polysomnography. Conducted overnight, usually in a sleep lab, this test monitors various physiological activities during sleep, including brain waves, eye movements, muscle activity, heart rhythm, and breathing patterns. The multi-faceted data collected helps to diagnose conditions like sleep apnea, restless leg syndrome, and narcolepsy.

For those who find the prospect of spending a night in a lab intimidating or impractical, home sleep tests have emerged as a convenient alternative. These portable devices measure some key indicators like oxygen levels, heart rate, and breathing patterns while the individual sleeps in their own bed. While not as detailed as polysomnography, home sleep tests are particularly useful for diagnosing sleep apnea and can be a stepping stone towards more specialized testing if needed.

Apart from these tests, there are also specialized evaluations for diagnosing less common sleep disorders. For instance, actigraphy involves wearing a wrist device that tracks movement and provides data on sleep-wake patterns over several days or weeks. This long-term monitoring can be particularly useful for understanding disturbances in circadian rhythms, often linked to jet lag and shift work sleep disorder.

Understanding when to see a specialist is crucial in the path to better sleep. If sleep issues persist despite improving sleep hygiene or if they significantly affect daytime functioning, consulting a sleep specialist is advised. Professionals may then recommend appropriate tests and interpret the results to tailor a treatment plan that addresses the specific underlying issues, ultimately guiding you towards more restful nights.

Polysomnography is a comprehensive and versatile diagnostic tool that's essential for understanding and treating various sleep disorders. This test, often conducted overnight in a sleep lab, allows specialists to monitor a wide array of physiological activities during sleep. If you've been grappling with chronic sleep issues, polysomnography might just be the key to unlocking a deeper understanding of what's happening when your head hits the pillow.

During a polysomnography test, multiple parameters are recorded simultaneously. These include brain wave activity (via EEG), eye movements (EOG), muscle activity (EMG), heart rhythm (ECG), and breathing patterns. It's an impressive array of data that collectively paints a detailed picture of your sleep architecture—essentially the various stages of sleep you cycle through each night. All this information helps identify disruptions in the sleep stages that could be causing distress.

One of the crucial roles of polysomnography is diagnosing sleep apnea, a common yet underdiagnosed condition where your breathing repeatedly stops and starts throughout the night. The breathing patterns and interruptions can be meticulously tracked with airflow sensors and oxygen saturation monitors. Understanding the severity and type of sleep apnea, whether obstructive or central, is vital for tailoring effective treatment methods like CPAP or lifestyle changes.

A polysomnography test isn't just about identifying sleep apnea. Conditions like periodic limb movement disorder (PLMD) and restless leg syndrome (RLS) are also often diagnosed through this study. By monitoring leg movements and muscle activity, doctors can detect abnormal movements that disrupt sleep. These conditions may not only affect your sleep quality but also have broader implications on your overall health and daytime functioning.

The brain's electrical activity recorded by EEG during polysomnography helps in identifying conditions like narcolepsy and parasomnias, which include unusual behaviors such as sleepwalking and night terrors. Variances in brain waves can indicate if you're suffering from fragmented sleep or if there are abnormalities in transitioning between different sleep stages.

Many might shy away from the idea of spending a night in a sleep lab, but it's worth noting that advancements in technology are making this process more comfortable and less intrusive. The benefit of obtaining accurate, highly detailed information about your sleep far outweighs the initial discomfort. The test setup is designed to closely mimic your home environment, making it easier to obtain results that reflect your usual sleep patterns.

An often overlooked aspect of polysomnography is its role in ensuring the efficacy of sleep disorder treatments. Once a treatment plan is in place, follow-up polysomnography might be recommended to confirm that the interventions are working as intended. Adjustments can be made based on this ongoing data to continually improve sleep quality.

Polysomnography is not typically the first step in diagnosing sleep issues; it usually comes into play after initial assessments indicate the need for in-depth analysis. Your doctor might suggest other preliminary steps like maintaining a sleep diary, using wrist actigraphy, or undergoing a less extensive home sleep test before opting for full-scale polysomnography.

However, the insights gained from this test extend beyond individual diagnoses. Data from polysomnographic studies contribute to a broader understanding of sleep disorders and their prevalence in the general population. It's an invaluable tool in sleep medicine that aids researchers in uncovering patterns and developing new treatment strategies.

If you're undergoing polysomnography, it may feel daunting, but the process is straightforward. After arriving at the sleep lab, you'll be prepped with sensors and leads placed on your scalp, face, chest, and limbs. These are all connected to a central data collection system. You'll then go to bed at your regular bedtime, and the real-time monitoring begins. Throughout the night, sleep technologists observe the data to ensure everything is recorded correctly.

As daunting as this might sound initially, remember that the goal is to help you achieve better sleep and, by extension, a higher quality of life. The intricate detailing of your sleep provided by polysomnography might just be what you need to understand those restless nights and work towards truly restorative sleep.

In conclusion, polysomnography stands as a cornerstone in the field of sleep medicine. It's not just about pinpointing problems; it's about constructing a route to sleep that revitalizes your body and mind. If sleep woes have been a shadow in your life, this test could illuminate the path to restful nights and energized days.

Home Sleep Tests are becoming an essential tool in modern sleep diagnostics. These tests offer a convenient and often more affordable alternative to traditional in-lab sleep studies, particularly when diagnosing conditions like sleep apnea. Instead of spending a night in a clinical setting, you can conduct the test in the comfort of your own home, allowing for a more natural reflection of your regular sleep patterns.

The process of using a home sleep test is straightforward and involves minimal discomfort. Typically, a kit is sent to your home, containing a variety of sensors and devices that you will setup yourself. These may include a nasal cannula to measure airflow, a finger clip to monitor oxygen saturation and heart rate, and belts to track respiratory effort. Instructions are often comprehensive and user-friendly to ensure the equipment is set up correctly.

One of the key benefits of home sleep tests is their accessibility. For those living in remote areas or who have mobility issues, this method eliminates the need to travel to a sleep lab. Furthermore, home sleep tests can significantly reduce the anxiety that some might feel about sleeping in an unfamiliar environment, leading to more natural sleep data and potentially more accurate results.

While traditional polysomnography remains the gold standard for diagnosing a wide range of sleep disorders, home sleep tests have particularly proven their worth in identifying moderate to severe obstructive sleep apnea (OSA). These tests focus on parameters essential for diagnosing OSA, like airflow, blood oxygen levels, and respiratory patterns. However, it's important to note that more complex sleep disorders may still require the comprehensive data provided by an in-lab study.

Another point to consider is the growing body of scientific literature supporting the efficacy of home sleep tests. Numerous studies have demonstrated their reliability and accuracy in diagnosing OSA, making them a credible first step in the diagnostic process. For health-conscious individuals and insomniacs, this represents an important advance in sleep medicine, making vital diagnostic tools more accessible and affordable.

Implementing a home sleep test is not without its challenges. The quality of the results can heavily depend on the user's ability to correctly setup the equipment. Therefore, adherence to the provided instructions is crucial. Some kits even come with virtual support to guide you through the process, ensuring that the sensors are placed correctly for accurate data collection.

Once the data is collected, it is usually sent back to a sleep specialist or a certified sleep technician for analysis. These professionals interpret the results and provide a diagnosis or recommendations for further action, which might include lifestyle changes, the use of CPAP (Continuous Positive Airway Pressure) machines, or even more detailed in-lab studies.

For many, the convenience and ease of home sleep tests outweigh the potential drawbacks. This method aligns well with the increasing trend towards remote healthcare solutions, empowering individuals to take charge of their health from the comfort of their own homes. It also offers a more flexible and less intrusive option for those with busy lifestyles or familial responsibilities that make overnight stays in a sleep lab challenging.

Given these advantages, it's not surprising that home sleep tests are gaining traction among both patients and healthcare providers. As technology improves, we can expect these tests to become even more sophisticated, potentially including additional metrics such as sleep stages and brain wave activity in the future. These advancements could further bridge the gap between home testing and traditional lab-based polysomnography, making sleep diagnostics more comprehensive and user-friendly.

In summary, **Home Sleep Tests** offer an innovative, accessible, and effective way to diagnose sleep disorders like obstructive sleep apnea. They provide a practical alternative to in-lab studies, especially for those who value convenience and comfort. As technology continues to evolve, these tests will likely become an even more integral part of sleep diagnostics, helping to enhance the quality of life for many individuals struggling with sleep issues.

When to See a Specialist

Sleep issues can often be managed with lifestyle adjustments, good sleep hygiene, and over-the-counter remedies. However, there comes a point when it's crucial to consult a specialist. Recognizing when to seek professional help can significantly impact your overall well-being and quality of life.

If you find yourself consistently struggling with sleep despite implementing best practices, it might be time to see a specialist. Occasional sleepless nights are normal, especially during stressful periods, but chronic sleep disturbances—those lasting more than a month—warrant professional attention. A qualified sleep specialist can help you identify underlying issues that might be disrupting your sleep, whether they're physical, psychological, or a combination of both.

One clear indicator that you need to consult a specialist is persistent daytime fatigue. If you're getting enough sleep but still feel exhausted during the day, this could signal an undiagnosed sleep disorder. Conditions like sleep apnea can cause interrupted sleep without you even realizing it, leading to chronic tiredness. A sleep specialist can administer tests such as polysomnography to diagnose these conditions accurately.

Frequent snoring, particularly if it's accompanied by gasping, choking sounds, or observed pauses in breathing, is another red flag. These symptoms often indicate sleep apnea, a potentially serious condition that requires medical intervention. Sleep apnea not only affects your sleep quality but is also linked to other health issues like high blood pressure and heart problems. A sleep specialist can recommend treatments ranging from lifestyle changes and CPAP machines to surgical options, depending on the severity of the condition.

Experiencing unusual behaviors during sleep, such as sleepwalking, night terrors, or acting out dreams aggressively, should also prompt you to seek specialist advice. These behaviors can be disruptive not only for you but also for anyone sharing your living space. A sleep specialist can help determine whether these actions are symptoms of a broader issue like REM sleep behavior disorder or another parasomnia, and suggest effective treatments.

In addition, if you have or suspect you have a mental health condition like depression, anxiety, or PTSD, and notice that it's impacting your sleep, consulting a specialist can be beneficial. Sleep disorders and mental health conditions often go hand-in-hand, exacerbating each other. A sleep specialist can work in conjunction with your mental health provider to develop a comprehensive treatment plan that addresses both aspects.

For individuals who have tried various insomnia treatments to no avail, a visit to a sleep specialist may reveal hidden factors contributing to sleeplessness. Sometimes, insomnia is not just a condition in itself but a symptom of another underlying issue such as restless leg syndrome or a circadian rhythm disorder. Accurate diagnosis and targeted treatment can make a significant difference.

Parents should consider seeing a specialist if their children have persistent sleep troubles. Sleep is crucial for a child's development, and ongoing issues can impact their mood, behavior, and learning capabilities. Pediatric sleep specialists can offer tailored advice and interventions that fit a child's unique needs.

In summary, while many sleep issues can be managed on your own, there are times when professional help is essential. Persistent insomnia, excessive daytime sleepiness, unusual nocturnal behaviors, and sleep disruption linked to mental health conditions are all valid reasons to consult a sleep specialist. Addressing these issues promptly can lead to better sleep and, by extension, a healthier, more fulfilling life.

Chapter 7:
Improving Sleep Hygiene

To truly improve your sleep hygiene, it's crucial to cultivate an environment and routine that support restful slumber. Start by designing a sleep-friendly bedroom: opt for blackout curtains to eliminate intrusive light, maintain a cool temperature, and consider investing in a comfortable mattress and pillows that suit your sleeping style. Lighting plays a pivotal role, so dim the lights as bedtime approaches to signal to your body that it's time to wind down. Establishing a consistent sleep routine is equally important; try to go to bed and wake up at the same times every day, even on weekends. This consistency helps regulate your internal clock, making it easier to fall asleep and wake up naturally. Simple steps like these can significantly enhance the quality of your sleep, setting the foundation for better overall health and well-being.

Creating a Sleep-friendly Environment

A sleep-friendly environment can significantly enhance the quality of your sleep by catering to your body's natural rhythms and minimizing disturbances. Start by ensuring your bedroom is a sanctuary dedicated to rest; keep it cool, dark, and quiet. Blackout curtains can eliminate intrusive light, and a white noise machine might help drown out unexpected sounds. Carefully consider your mattress and pillows, aiming for comfort and support that suit your needs. Additionally, removing electronic devices or at least limiting their use before bedtime can mitigate the effects of disruptive blue light. By curating a space that fosters relaxation and comfort, you'll set the stage for a restful and rejuvenating night's sleep.

Ideal Bedroom Setup

The environment in which you sleep plays a crucial role in the quality of your rest. Crafting an ideal bedroom setup isn't just a matter of aesthetics; everything from the temperature to the type of mattress you choose can impact how well you sleep. To enhance sleep quality and improve overall well-being, making mindful adjustments to your bedroom is essential.

First and foremost, consider the *temperature* of your bedroom. Research indicates that a cooler room, generally between 60 and 67 degrees Fahrenheit, facilitates better sleep. When it's too warm, your body may struggle to lower its internal temperature enough to initiate sleep, leading to restlessness. On the other hand, a room that's too cold can prevent you from falling asleep quickly. Experiment with your thermostat to find the setting that feels just right for you.

Next is **lighting.** The lighting in your bedroom significantly influences your sleep-wake cycle, also known as the circadian rhythm. Exposure to bright artificial light, especially blue light emitted from screens, can inhibit the production of melatonin, a hormone responsible for sleep. Consider using blackout curtains to block any external light and opt for dim, warm lighting in the evening hours. This shift from bright to dim light helps signal your body that it's time to wind down.

Another vital aspect is the *bed and bedding.* Your mattress should support your body without causing any pressure points, which means it shouldn't be too firm or too soft. Material preferences vary, but memory foam and latex mattresses are excellent for contouring to your body's shape. Additionally, invest in high-quality pillows that keep your neck aligned with your spine. For those who suffer from allergies, hypoallergenic pillows and mattress covers can make a world of difference.

Let's not overlook **sound.** Noise can be a massive disruptor of sleep. If you live in a noisy environment, consider soundproofing your room or using a white noise machine. These devices can mask disruptive sounds and create a consistent auditory environment that can help you relax. Similarly, earplugs can offer a quick and effective solution to block out unwanted noise. Soft, calming sounds like those from nature can also set the stage for a restful night's sleep.

A crucial yet often ignored element is *clutter.* A clean, clutter-free bedroom promotes relaxation and reduces stress. When you're surrounded by mess, it can subconsciously keep your mind engaged, making it harder to relax. Make it a habit to keep your bedroom tidy. Incorporate storage solutions that can keep personal items organized and out of sight, creating a peaceful environment dedicated to sleep.

Meanwhile, consider incorporating **scents** that promote relaxation. Aromatherapy can be a powerful tool in your sleep arsenal. Essential oils like lavender and chamomile are known for their calming effects. Diffusing these scents in your bedroom or adding a few drops to your pillow can help induce a state of relaxation. Be cautious with the concentration, as overly strong scents can sometimes be counterproductive.

Moreover, think about the *placement* of furniture. The layout of your bedroom should contribute to a sense of tranquility. Position your bed so you can see the door but avoid being directly in line with it. This layout can instill a sense of security and ease. Keep nightstands within arm's reach for convenience but avoid overloading them with gadgets and other distractions.

Speaking of **gadgets,** the ideal bedroom should be a tech-free zone. The presence of smartphones, tablets, and televisions can interfere with your ability to sleep. These devices emit blue light, which can trick your brain into thinking it's still daytime, thus delaying sleep. Establish a technology curfew to disconnect from screens at least an hour before bedtime. This habit can help your mind transition into a state conducive to sleep.

Plants can also contribute to an ideal bedroom setup. *Indoor greenery* not only beautifies your space but also serves as a natural air purifier. Plants like snake plants and peace lilies can improve air quality, creating a more oxygenated and serene sleeping environment. Just be sure to choose low-maintenance plants that won't require too much attention.

Don't forget about **personalizing** your space to reflect calm. Colors play a psychological role in how we perceive our environment. Calming hues such as soft blues, greens, and neutrals can foster relaxation. Avoid overly stimulating colors like bright reds or yellows, which can have the opposite effect. Personal touches like framed photographs, meaningful artwork, or a favorite blanket can create a sense of comfort and security, making your bedroom a shrine of rest and rejuvenation.

An often overlooked but critical element is your *flooring*. While it might not seem directly related to sleep, stepping onto a cold or rough floor can jolt you awake. Soft, warm carpets or strategically placed area rugs can provide comfort underfoot and help ease your transition in and out of bed.

Finally, let's not underestimate the importance of **routine**. The best-designed bedroom still requires consistent sleep habits to be effective. This includes maintaining a regular sleep schedule and creating a bedtime ritual. Such routines reinforce your body's internal clock, making it easier to fall and stay asleep. Whether it's a warm bath, reading a book, or practicing some light stretches, finding what relaxes you before bed can make your ideal bedroom setup all the more effective.

Achieving the perfect bedroom environment isn't a one-size-fits-all endeavor. It requires personal tweaks and patience to discover what works best for you. By concentrating on these key aspects of your bedroom setup, you'll be well on your way to transforming your sleep quality—because when it comes to sleep, your environment truly makes a world of difference.

The Role of Lighting is a critical aspect of creating a sleep-friendly environment that often gets overlooked. The kind of lighting you expose yourself to throughout the day and night can significantly impact your ability to fall asleep and the quality of sleep you receive. This isn't just about whether the lights are on or off when you go to bed, but also about the type, intensity, and timing of light exposure.

Our bodies are incredibly sensitive to light. Light influences our circadian rhythms, the internal clocks that regulate sleep-wake cycles. In natural environments, these rhythms follow the rising and setting of the sun. However, in today's world, artificial lighting can disturb this natural order. Exposing yourself to bright, blue-rich light during the evening can trick your brain into thinking it's still daytime, making it harder for you to wind down and fall asleep.

Consider the impact of blue light, which is emitted by screens on phones, tablets, and computers, as well as energy-efficient LED and fluorescent lighting. Blue light has a short wavelength that can delay the release of sleep-inducing melatonin, disrupt your circadian rhythm, and make it more challenging to fall asleep. If you can't avoid screens before bedtime, using blue light filters or apps that adjust your screen's color temperature can be a helpful mitigation strategy.

In contrast, exposure to natural light during the morning and daytime can be enormously beneficial. Daylight helps to synchronize your internal clock with the outside world, promoting alertness during the day and sleepiness in the evening. Try to spend time outdoors in natural light, especially in the morning. If that's not feasible, consider full-spectrum light therapy lamps as a substitute. These lamps can simulate daylight and have been shown to improve mood and sleep quality.

During the evening, it's wise to create a dim, warm-lit environment. Soft, warm-toned lights create a conducive atmosphere for relaxation as they mimic the qualities of twilight, signaling to your body that it's almost time for sleep. Consider using dimmers or lamps with adjustable brightness levels. This doesn't just apply to your bedroom but also to other areas you spend time in during the evening, like the living room or bathroom.

The strategic use of lighting can transform your bedroom into a sanctuary for sleep. Avoid overhead lights and opt for bedside lamps with shades that diffuse light softly. Red or amber light bulbs are less likely to interfere with melatonin production than regular white or blue-toned bulbs. A blackout curtain can also be a valuable investment to block out external light sources, such as streetlights, that might sneak into your room during the night.

While indoor lighting designs may be functional or aesthetic, it's crucial to consider their biological effects. Bright, overhead lighting during the night can be disruptive. Instead, aim for lower, floor or table-level lighting options that cast a gentle glow. This approach not only helps relax your eyes but also maintains your circadian alignment, making it easier to fall asleep once you hit the pillow.

Lighting isn't just about electric lights, either. Candles can be an excellent, low-light alternative for the evening hours. They provide soft, warm light without any blue wavelengths, making them ideal for pre-sleep routines. However, ensure safety by never leaving them unattended or in easily flammable areas.

Balancing the quality and timing of light exposure doesn't stop once you're in bed. If you're someone who often needs to get up during the night, consider installing motion sensor nightlights in hallways or bathrooms. These lights should be very soft and low to minimize sleep disturbances.

Understanding the role of lighting can make a world of difference in improving your sleep quality. It's not just about quantity of sleep; it's about aligning your sleep environment with your body's natural rhythms. This simple, yet powerful, lifestyle adjustment can enhance your overall well-being, making you feel more rested and alert during the day. In the grand scheme of sleep hygiene, lighting is a cornerstone element that sets the stage for a night of restorative sleep.

Establishing a Sleep Routine

One of the core components of improving sleep hygiene is establishing a consistent sleep routine. Consistency is key when it comes to regulating your body's internal clock, also known as the circadian rhythm. By going to bed and waking up at the same times each day, even on weekends, you'll train your body to expect sleep at specific times, ultimately making it easier to fall asleep and wake up naturally.

Begin by setting a fixed bedtime that allows for 7-9 hours of sleep, depending on your personal needs. Consider working backward from your desired wake-up time. For instance, if you need to wake up at 6 AM, aim to be in bed by 10 PM. Moreover, keep in mind that your wind-down routine should start at least 30 minutes before your set bedtime. This period should be devoted to relaxing activities that signal to your body it's time to prepare for sleep.

During this wind-down period, engage in calm activities such as reading a book, taking a warm bath, or practicing gentle stretching. Avoid stimulating activities like intense exercise, watching TV, or browsing on your phone, as these can interfere with your body's ability to relax. Lowering the lights in your home can also help cue your brain that it's time to start winding down.

If you live in an environment where noise can disrupt your sleep, consider incorporating white noise or other soothing sounds into your bedtime routine. Devices or apps that produce these sounds can help mask disruptive external noises, creating a more conducive sleep environment.

In addition to your pre-sleep routine, what you do throughout the day also greatly impacts your ability to establish a healthy sleep routine. Exposure to natural daylight, particularly in the morning, can help regulate your circadian rhythm. Spending time outside or near a window first thing in the morning can signal to your brain that it's time to be awake, making it easier to maintain a consistent sleep schedule.

As your day progresses, it's crucial to manage your intake of stimulants such as caffeine. Consuming caffeine even up to six hours before bedtime can impair your ability to fall asleep. Hence, avoiding caffeine in the afternoon and evening is an essential part of establishing a successful sleep routine.

Consistency is particularly challenging but necessary when it comes to weekends. The temptation to sleep in can disrupt your established routine, leading to what some experts call "social jetlag." To avoid this, try to wake up at your usual time and follow the same bedtime practices, making only minor adjustments if needed.

For many, the struggle isn't just going to bed at the right time but also ensuring the quality of sleep is optimal. Establishing a sleep routine is about creating rituals that promote relaxation and signal to your body that the day is winding down. It's a methodical way of ensuring that both the quantity and quality of your sleep are maximized, thereby improving your overall well-being and daily functionality.

Lastly, be patient with yourself as you adapt to a new sleep routine. It may take several weeks for your body to adjust fully. If you continue to experience difficulties despite following these guidelines, it may be beneficial to consult a sleep specialist to rule out any underlying issues or to get personalized advice.

Chapter 8:
Nutrition and Sleep

Our bodies don't just shut off when we sleep, and the food we consume plays a significant role in how well we rest. Certain nutrients, like tryptophan, found in turkey and dairy, contribute to the production of serotonin and melatonin, which help regulate our sleep cycles. However, not all foods are beneficial; caffeine and alcohol, though often used as sleep aids, can disrupt sleep patterns and diminish overall sleep quality. Incorporating supplements and herbs, such as magnesium and valerian root, into your diet can also support better sleep. By thoughtfully choosing what we eat and drink, we can enhance our sleep quality and, in turn, our overall well-being.

Foods That Promote Sleep

Nutrition plays a vital role in sleep quality, and certain foods can significantly aid in promoting better rest. Incorporating sleep-friendly foods into your diet can be a natural and effective way to improve your overall sleep hygiene. But what specifically should we be eating to help us drift off more easily?

First, let's talk about tryptophan. Tryptophan is an amino acid found in various foods that's essential for the production of serotonin, which in turn is converted into melatonin, the sleep-regulating hormone. Foods rich in tryptophan include turkey, chicken, eggs, nuts, seeds, and dairy products. Incorporating these items into your evening meal could give your body the building blocks it needs to produce melatonin, making it easier to fall asleep.

Complex carbohydrates also play a critical role. Foods like whole grains, oats, and brown rice have a slower digestive rate, which helps maintain stable blood sugar levels during the night. Combining complex carbs with tryptophan-rich foods can be even more effective. For example, a turkey sandwich on whole grain bread or a bowl of oats with a sprinkle of nuts can be excellent choices.

It's also beneficial to include foods that are high in certain minerals. Magnesium and potassium are well-known for their muscle-relaxing properties, which can aid in reducing muscle cramps and tension that might otherwise disturb your sleep. Bananas, avocados, leafy greens, and nuts like almonds are particularly good sources of these minerals.

Herbal teas, such as chamomile and valerian root tea, have also been traditionally used to promote sleep. Chamomile contains an antioxidant called apigenin, which binds to specific receptors in your brain that may decrease anxiety and initiate sleep. Valerian root, on the other hand, has compounds that boost gamma-aminobutyric acid (GABA) levels in the brain, promoting a relaxing effect.

Timing is crucial as well. Avoid large meals before bedtime as they can cause discomfort and indigestion. A light snack with the right combination of nutrients, like a small banana with a handful of almonds or a piece of whole grain toast with peanut butter, can be both satisfying and sleep-promoting. Balancing your diet to include these sleep-friendly foods can be a cornerstone of improving your sleep quality and overall well-being.

The Role of Tryptophan in your diet is crucial for getting that much-needed shut-eye. Tryptophan is an essential amino acid, meaning your body can't produce it on its own, so you have to get it from the foods you eat. Once consumed, tryptophan travels to the brain, where it's converted into serotonin, a neurotransmitter that promotes relaxation. Serotonin is then converted to melatonin, the hormone most closely associated with sleep-wake cycles.

You're probably familiar with common sources of tryptophan like turkey, which is often blamed for post-Thanksgiving dinner drowsiness. But other foods high in tryptophan include nuts, seeds, tofu, cheese, red meat, chicken, fish, oats, beans, and eggs. Including these foods in your evening meal could help your body produce the serotonin and melatonin necessary for restful sleep.

What's fascinating is that tryptophan's real magic happens when combined with carbohydrates. Carbs cause your body to release insulin, which helps remove competing amino acids from your blood. This process allows tryptophan to cross the blood-brain barrier more efficiently. That's why a late-night snack of cheese and crackers or yogurt and fruit can be a double whammy for promoting sleep.

Yet, it's not just about the food you eat. Timing is essential. Consuming tryptophan-rich foods a few hours before bed can maximize its potential benefits. That's because it takes some time for your body to digest the food, extract the tryptophan, and convert it into melatonin. So, a well-timed dinner or evening snack can set the stage for a good night's sleep.

Now, let's get into why tryptophan could be especially beneficial for insomniacs. Research suggests that individuals with insomnia often struggle with low levels of serotonin and melatonin. By ensuring a steady intake of tryptophan, insomniacs may be able to naturally boost their serotonin and melatonin levels, thus improving their ability to both fall asleep and stay asleep. This isn't a cure-all, but it's a natural approach to potentially alleviating some sleep issues.

Bear in mind, though, that not everyone will experience a dramatic difference by simply upping their tryptophan intake. Sleep is influenced by a myriad of factors, from stress levels to sleep environment, and tryptophan is just one piece of the puzzle. However, it's an easily adjustable one, accessible through everyday dietary choices.

Some studies even highlight that tryptophan supplements can be effective in improving sleep latency and efficiency. But before opting for supplements, it's always best to consult with a healthcare provider. That way, you can avoid any potential interactions with other medications or unintended side effects.

As you navigate your journey to improved sleep, consider how incorporating tryptophan-rich foods into your diet can be a simple yet effective tool. The beauty lies in its simplicity and natural origin, making it an attractive option for those wary of pharmaceutical interventions. Remember, the path to better sleep often involves several small adjustments, and tweaking your diet to include more tryptophan could be one of those beneficial steps.

The Impact of Caffeine and Alcohol on sleep cannot be understated, especially when you're trying to enhance sleep quality and overall well-being. Both substances have significant but distinct effects on your body's ability to transition into restful states. Understanding their mechanisms can help you make more informed choices about when and how to consume these beverages.

Caffeine, found in coffee, tea, soft drinks, and many other products, is the world's most widely used psychoactive substance. It works by blocking the action of adenosine, a neurotransmitter that facilitates sleep and relaxation. While this might be just what you need for a mid-day energy boost, it can spell trouble for your sleep later. The half-life of caffeine can vary, but it generally lasts about five to six hours. This means that a cup of coffee consumed in the late afternoon could still be affecting your nervous system come bedtime. For those particularly sensitive to caffeine, even a morning dose can linger and impair nighttime sleep.

Choosing your timing wisely is crucial. Many experts recommend avoiding caffeine at least six hours before bedtime. If you struggle with falling asleep, it may be beneficial to create a longer buffer period. However, it's not just about the clock; personal sensitivity plays a significant role. Some people metabolize caffeine quickly, while others do so more slowly. Recognizing your own body's response is key to managing its impact.

On the flip side, alcohol presents a different kind of challenge. Although it might initially seem to promote sleepiness, making people think of it as a nightcap, alcohol disrupts the sleep cycle. It can help you fall asleep faster, but the quality of that sleep is compromised. Alcohol increases the likelihood of waking up in the middle of the night and reduces the restorative REM sleep stages that are crucial for cognitive function and emotional regulation.

Ingesting alcohol before bed can also worsen snoring and sleep apnea symptoms. Even moderate levels of alcohol consumption have been shown to elevate these risks. Additionally, the diuretic effects of alcohol might lead to more nighttime awakenings to use the bathroom, further interrupting your sleep and leaving you less refreshed in the morning.

The combined effects of caffeine and alcohol can be particularly detrimental. Consuming these substances too close together can create a confusing cocktail of signals for your body to process. On one end, caffeine keeps you wired, and on the other, alcohol sets you up for fragmented and poor-quality sleep. It's a double whammy that can sabotage even the best efforts to maintain a healthy sleep routine.

So, what can you do to mitigate these effects? First and foremost, awareness is your best tool. Keep track of your intake, log the times you consume caffeine and alcohol, and note how it affects your sleep on those days. This can help you identify patterns and make necessary adjustments. Opt for herbal teas or other non-caffeinated and non-alcoholic beverages as bedtime approaches. You might discover that simply switching to these alternatives significantly improves your sleep quality.

Furthermore, understand that you don't have to give up caffeine or alcohol entirely to enjoy better sleep. Moderation and timing are key. Experiment with reducing your intake gradually and observe the impact on your sleep. Small changes can make a big difference, such as limiting caffeine to the morning hours or enjoying that glass of wine earlier in the evening rather than just before bed.

In summary, caffeine and alcohol have complex and often opposing effects on sleep quality. Knowing how and when to consume these substances can be a powerful strategy in your quest for better sleep. Be mindful, experiment thoughtfully, and remember that improving your sleep is an ongoing journey marked by gradual, sustainable habits that lead to healthier days and more restful nights.

Supplements and Herbs play a pivotal role in the delicate dance of achieving restful sleep. As research delves deeper into the science of sleep, it uncovers a myriad of natural remedies that may ease the path to slumber. Supplements and herbs target various aspects of sleep physiology, potentially helping you fall asleep quicker, stay asleep longer, and wake up feeling refreshed. But it's crucial to understand that these remedies don't work in isolation; they're most effective when combined with good sleep hygiene practices.

One of the most well-researched natural sleep aids is melatonin, a hormone produced by the pineal gland in the brain. Melatonin levels rise in the evening, signaling to your body that it's time to wind down and prepare for sleep. Supplements containing melatonin can be particularly beneficial for people suffering from disrupted sleep cycles, such as shift workers or those experiencing jet lag. However, it's essential to use melatonin cautiously and consult with healthcare providers since overuse or incorrect timing can lead to grogginess or disrupt natural melatonin production.

Herbal remedies also offer a range of options for enhancing sleep quality. Valerian root, for instance, has been used for centuries to treat insomnia and anxiety. This herb may work by increasing levels of gamma-aminobutyric acid (GABA) in the brain, which has a calming effect on the nervous system. Unlike prescription medications, valerian is generally well-tolerated with fewer side effects, though it may take a few weeks of consistent use to notice significant benefits.

Another popular herb for sleep is chamomile, often consumed as a tea. Chamomile contains antioxidants like apigenin, which binds to specific receptors in the brain that may promote sleepiness and reduce insomnia symptoms. The soothing ritual of sipping chamomile tea can also act as a signal to your body that it's time to relax, adding a psychological layer of benefit.

Magnesium is a mineral that plays a crucial role in sleep regulation. It helps activate the parasympathetic nervous system, which is responsible for making you feel calm and relaxed. Magnesium supplements are widely available, and some studies suggest they may improve sleep quality, especially for individuals with low magnesium levels. Incorporating magnesium-rich foods into your diet, like almonds, spinach, and yogurt, can also contribute to better sleep health.

L-theanine, an amino acid found predominantly in green tea, is another supplement worth considering. Unlike caffeine, L-theanine promotes relaxation without drowsiness. It works by modulating levels of neurotransmitters like serotonin and dopamine, which can help alleviate stress and anxiety, paving the way for a more restful night's sleep. Combining L-theanine with a small dose of melatonin might provide a synergistic effect, although it's always best to discuss such combinations with a healthcare professional.

Lavender, whether used in essential oils, teas, or supplements, is famous for its calming and sedative properties. The aroma of lavender has been shown to decrease heart rate and blood pressure, creating an optimal environment for sleep. Applying a few drops of lavender oil to your pillow or using it in a bedtime bath can be an effective way to utilize its sleep-promoting benefits.

It's important to approach supplements and herbs with a sense of curiosity and caution. While natural remedies can offer significant benefits, they are not one-size-fits-all solutions. Individual responses can vary widely, and what works wonders for one person might not be effective for another. Prioritize consulting with healthcare providers to tailor a regimen that fits your specific needs and ensures there are no contraindications with other medications you may be taking.

In summary, combining these natural remedies with other aspects of sleep hygiene can create a holistic approach to improving sleep quality. Supplements and herbs like melatonin, valerian root, chamomile, magnesium, L-theanine, and lavender offer various mechanisms to promote relaxation and manage sleep disorders. By incorporating these into your nightly routine, you can transform the elusive quest for restful sleep into a more attainable reality.

Chapter 9:
Exercise and Sleep

Physical activity and sleep are deeply intertwined, each influencing the other in profound ways. Regular exercise has been shown to enhance sleep quality by reducing the time it takes to fall asleep and increasing the duration of deep sleep stages. However, it's not just about doing any exercise; specific types such as aerobic activities, strength training, and mind-body exercises like yoga and Tai Chi can be particularly beneficial. Timing is also critical—exercising too close to bedtime may have the opposite effect, elevating heart rate and body temperature and making it difficult to wind down. For optimal results, aim for moderate-intensity workouts earlier in the day. By finding the right balance and routine that works for you, you can harness the power of exercise to not only improve your overall health but also enjoy more restful and rejuvenating sleep.

How Physical Activity Affects Sleep

Engaging in regular physical activity can work wonders for your sleep quality. Exercise increases the time spent in deep sleep—the most physically restorative sleep phase. Deep sleep is essential for immune function, cardiovascular health, and managing stress and anxiety. When you exercise, your body's temperature rises, and the post-exercise drop in temperature can facilitate the onset of sleep, leading to a more restful night.

Different forms of exercise impact sleep in various ways. Aerobic activities like running, cycling, and swimming are particularly effective at improving sleep. These workouts elevate your heart rate and release endorphins, which can improve mood and make it easier to fall asleep. Resistance training and high-intensity interval training (HIIT) also contribute to better sleep, although they might be more stimulating if done too close to bedtime.

Alongside the physiological benefits, exercise also helps reduce symptoms of sleep disorders such as insomnia and sleep apnea. By decreasing feelings of anxiety and depression, regular physical activity can break the cycle of stress-induced sleeplessness. For insomniacs, establishing a consistent exercise routine can serve as a natural remedy, helping reset the body's internal clock and promote a regular sleep-wake cycle.

Moreover, the timing of your workout plays a significant role. Morning and afternoon workouts are generally more beneficial for sleep than late-evening sessions. Engaging in physical activities earlier in the day helps regulate circadian rhythms, making it easier to fall asleep at night. If you prefer evening workouts, low-impact activities like yoga or gentle stretching might be more suitable as they won't leave you too energized to wind down.

Incorporating physical activity into your daily routine doesn't have to be arduous or time-consuming. Even short bursts of movement, such as walking or taking the stairs, can make a notable difference in your sleep quality. The key is consistency and finding an exercise you enjoy, ensuring it becomes a seamless part of your daily life.

Best Types of Exercise for Sleep

When it comes to enhancing sleep quality, exercise can be a powerful tool. So, what types of exercise are best suited for improving sleep? The answer isn't one-size-fits-all, but rather it depends on various factors, including personal preferences, fitness levels, and specific sleep challenges.

Aerobic exercise, often referred to as "cardio," is one of the most researched and proven forms of physical activity for sleep improvement. Activities such as jogging, swimming, cycling, or even walking at a brisk pace can help increase the amount of deep sleep you get. Deep sleep, also known as slow-wave sleep, is essential for feeling refreshed and rejuvenated. Studies have shown that individuals who engage in regular aerobic exercises fall asleep faster and spend more time in deep sleep phases compared to those who do not exercise.

Strength training, which includes weight lifting and resistance exercises, also offers benefits for sleep quality. While strength training primarily focuses on building muscle mass, it positively affects your sleep cycle by helping in the release of endorphins, which can reduce stress levels—a common barrier to good sleep. Moreover, engaging in strength training can contribute to overall physical fatigue, making it easier to fall asleep at night.

Mind-body exercises like yoga and tai chi are particularly effective for those who suffer from insomnia or light sleep. These exercises combine physical movement with mental focus and controlled breathing, which helps to relax the mind and body. Yoga, in particular, has been shown to improve various aspects of sleep, such as sleep efficiency, total sleep time, and the number of awakenings throughout the night. Practices like restorative yoga and yoga nidra can be especially beneficial for winding down before bed.

High-Intensity Interval Training (HIIT) has been gaining popularity for its efficiency in burning calories and improving cardiovascular health. HIIT involves short bursts of intense exercise followed by rest or low-intensity exercise. However, it's crucial to time these workouts correctly. Doing HIIT workouts too close to bedtime may lead to increased heart rate and adrenaline levels, which could interfere with your ability to fall asleep. Conducting these workouts earlier in the day can help you benefit from the elevated mood and reduced stress levels that HIIT provides, without compromising your sleep.

Stretching and flexibility exercises might not be the first things that come to mind when thinking about exercise for sleep, but they play a crucial role. Stretching helps to release muscle tension accumulated throughout the day, promotes relaxation, and prepares your body for rest. Simple stretches targeting major muscle groups can be easily incorporated into a bedtime routine to enhance sleep quality.

Swimming stands out as a full-body workout that can be incredibly beneficial for sleep. The buoyancy of water relieves physical stress on the body, making it an excellent option for people with joint issues or arthritis. The rhythmic nature of swimming laps can be meditative, easing mental stress, which is often a barrier to good sleep. Additionally, the physical exertion from swimming can make falling asleep easier and lead to more profound, uninterrupted sleep cycles.

Outdoor exercises, such as hiking or trail running, provide added benefits from exposure to natural light. Sunlight helps regulate your circadian rhythm, the body's natural sleep-wake cycle. Morning outdoor activities have the dual benefit of physical exercise and natural sunlight exposure, which can be particularly effective in helping you fall asleep faster at night.

Group fitness classes like Zumba or spin classes offer the added benefit of social interaction, which can enhance mood and contribute to lower stress levels. The collective energy and motivation found in group settings can make workouts more enjoyable, potentially leading to a more consistent exercise routine. Consistency is key; regular exercise is more effective for sleep improvement than sporadic activity.

Lastly, consider incorporating low-impact exercises such as Pilates into your routine. Pilates focuses on core strength, flexibility, and mindful movement, contributing to reduced muscle tension and improved overall relaxation. These benefits can translate directly into better sleep quality.

In summary, the best types of exercise for sleep vary depending on individual preferences and specific sleep issues. Aerobic activities, strength training, mind-body exercises like yoga, and even stretching all offer unique benefits that contribute to better sleep. The key is to find an exercise routine that you enjoy and can stick to consistently. The timing of exercise also plays a role; while vigorous workouts may need to be completed earlier in the day, relaxing activities such as yoga or stretching can be incorporated into your bedtime routine for immediate sleep benefits.

So, lace up those sneakers, roll out that yoga mat, or dive into the pool—your journey to better sleep can be as enjoyable and diverse as the exercises you choose.

Timing Your Workouts is crucial when it comes to optimizing both your exercise performance and your sleep quality. When you exercise can have a significant impact on your sleep architecture, the different stages of sleep you cycle through at night. It's not just about how much you're working out, but when you're doing it that counts.

Research shows that moderate to vigorous exercise can improve sleep efficiency, which is the ratio of time spent asleep to the total time spent in bed. However, if you don't time your physical activity appropriately, you may inadvertently worsen your sleep issues rather than alleviate them. For instance, working out too close to bedtime can elevate your heart rate, body temperature, and adrenaline levels, making it hard to fall asleep.

It's generally advisable to finish strenuous workouts at least three hours before you plan to go to bed. Doing so allows your body ample time to return to its baseline state. During exercise, your body's core temperature rises; upon cessation, it begins to cool down, mirroring the natural drop in body temperature that facilitates the onset of sleep. Consistent exercise performed earlier in the day has also been shown to help stabilize your circadian rhythm, the internal clock that dictates your sleep-wake cycle.

Morning workouts can be especially beneficial if you're trying to reset an erratic sleep schedule. Exercising in the morning exposes you to natural light, which can reinforce your circadian rhythm. Morning light exposure signals to your brain that it's time to wake up and be alert, making it easier for you to feel sleepy when night comes. If you're dealing with insomnia, this could be a game-changer.

On the other hand, afternoon workouts have their own set of advantages. During this time, your body temperature is already starting to rise naturally, which might make it easier for you to perform at your physical peak. Afternoon exercise has also been found to lead to deeper sleep stages and quicker sleep onset, provided you don't exercise too late in the day. The key lies in finding the right balance and observing how your body reacts.

Even evening workouts have their place and can be beneficial for some. Low-intensity exercises such as yoga or light stretching can help promote relaxation and trigger the release of melatonin, the hormone responsible for sleep. If you're not a fan of early morning or afternoon workouts, consider engaging in these more calming forms of exercise to help wind down from the day's stresses.

Your individual lifestyle and preferences will also dictate the best time for you to exercise. Some people naturally feel more energetic and motivated at different times of the day. Listen to your body and carefully note how your workout schedule influences your sleep patterns. Using a sleep journal or a sleep tracking device can help you identify trends and make more informed decisions.

It's also important to be mindful of the type of exercise you're engaging in. High-intensity interval training (HIIT) or heavy weightlifting sessions are best reserved for earlier in the day due to their stimulating nature. Sessions that prioritize cardio or moderate strength training can be more flexible in terms of timing but still require some buffer time before bed. On the other end of the spectrum, activities that focus on mindfulness and breath control, such as yoga or tai chi, are excellent choices for evening workouts.

When considering timing, don't overlook the consistency factor. Regular physical activity at a similar time each day helps establish a routine, which can be comforting to your body's internal clock. Whether it's a morning jog or an afternoon swim, making it a staple in your daily schedule can offer long-term benefits for both your fitness and sleep health.

Understanding the interplay between exercise and sleep extends beyond the gym. Nutrition, hydration, and overall lifestyle choices all contribute to how well you rest at night. Avoid stimulating foods and beverages like caffeine close to your planned workout time and subsequent bedtime. Maintaining a balanced diet rich in sleep-promoting nutrients like magnesium and tryptophan can further enhance the benefits of a well-timed workout regimen.

By paying attention to when you work out, you can create a synergy between your physical activity and sleep patterns. Getting this timing right can make a noticeable difference in how quickly you fall asleep, how deeply you sleep, and how refreshed you feel in the morning. Incorporating these strategies will help you not only achieve your fitness goals but also enjoy the restorative sleep that your body needs to maintain overall well-being.

Remember, there's no one-size-fits-all approach. What works wonders for one person might not have the same effect on another. Experiment with different times, track your results, and find the optimal workout schedule that helps you achieve the best sleep of your life. In doing so, you'll be taking a critical step towards better health and improved quality of life.

Mind-Body Exercises Like Yoga and Tai Chi are ancient practices that combine physical postures, breathing techniques, and meditation. These exercises don't just aim for physical fitness but target holistic well-being by creating harmony between the mind and body. Both yoga and Tai Chi have been extensively studied for their various health benefits, and their positive impact on sleep quality is particularly noteworthy.

Yoga originated in ancient India and encompasses a wide range of styles and practices. Whether you're doing a vigorous Vinyasa flow or a gentle Hatha sequence, yoga actively encourages you to focus on your breath and movement. This focus creates a meditative state that can significantly reduce stress levels. Stress, as we know, is a major contributor to sleep problems. Therefore, by lowering cortisol levels and promoting relaxation, yoga naturally aids in better sleep.

One of the most effective ways yoga helps improve sleep is through the practice of specific poses, or asanas, designed to promote relaxation. Poses like Supta Baddha Konasana (Reclining Bound Angle Pose), Viparita Karani (Legs-Up-The-Wall Pose), and Savasana (Corpse Pose) are particularly conducive to a restful state. When practiced regularly before bedtime, these poses can prepare your body and mind for sleep, making it easier to drift off and stay asleep. Moreover, these poses often involve slow, deep breathing, which activates the parasympathetic nervous system, promoting a state of calm.

Similarly, Tai Chi is a Chinese martial art characterized by slow, deliberate movements and deep breathing. It's often described as "meditation in motion." This practice is rooted in the principles of yin and yang, aiming to balance and harmonize the body's energy. Tai Chi's gentle, flowing movements can help alleviate stress and anxiety, two major hurdles in achieving good sleep.

The benefits of Tai Chi for sleep are backed by scientific research. Multiple studies have shown that regular practice of Tai Chi can improve sleep quality among older adults, individuals with chronic conditions, and even those experiencing the stresses of everyday life. Like yoga, Tai Chi encourages deep, rhythmic breathing, which can oxygenate the brain and prepare the body for restful sleep.

Another advantage of both yoga and Tai Chi is their adaptability. You don't need any special equipment or a gym membership. These exercises can be performed in the comfort of your home, making them accessible to everyone. This ease of access means you can seamlessly incorporate them into your daily routine, whether it's a quick morning sequence or a restful evening practice. Consistency is key to reaping the sleep benefits these practices offer.

Breathing techniques are an essential component of both yoga and Tai Chi. Practices like Pranayama in yoga or deep abdominal breathing in Tai Chi can immediately impact your autonomic nervous system. These breathing exercises promote a sense of calm and focus, making it easier to transition into sleep. Additionally, both practices encourage 'diaphragmatic breathing,' which reduces the body's stress response and can help with quicker sleep onset and better sleep maintenance.

It's also worth mentioning the role of mindfulness in these practices. Mindfulness, essentially the practice of being present in the moment, is a critical component of yoga and Tai Chi. By focusing your mind on the present, you can let go of the stress and worries that keep you awake at night. This mindful state activates the body's relaxation response, which is essential for falling and staying asleep.

The meditative aspect of these practices shouldn't be overlooked either. Meditation has been shown to increase melatonin levels, a hormone that regulates sleep-wake cycles. While dedicated meditation sessions can be highly beneficial, the meditative quality of mind-body exercises like yoga and Tai Chi means you're essentially killing two birds with one stone—getting the benefits of physical exercise and the restorative effects of meditation simultaneously.

You might wonder if one is better than the other for improving sleep. The truth is, both yoga and Tai Chi offer unique benefits, and the best choice depends on your personal preferences and physical condition. Some people might prefer the dynamic flow and variety of yoga, while others might be drawn to the slow, gentle nature of Tai Chi. The key is to find a practice that resonates with you and stick with it. Consistency and regularity will yield the best results.

To conclude, integrating mind-body exercises like yoga and Tai Chi into your daily routine can be a transformative step towards enhancing your sleep quality and overall well-being. These practices offer not just physical benefits but also mental and emotional relief. In a fast-paced world filled with stressors, giving yourself the gift of a few moments of peace and alignment can make a world of difference in your sleep and life quality. So, unroll your yoga mat or find a quiet spot to practice Tai Chi and take the first step toward a more restful night.

Chapter 10:
Stress Management and
Relaxation Techniques

Integrating effective stress management and relaxation techniques into your daily routine can be a game-changer for enhancing sleep quality. Stress is a major deterrent to a good night's sleep, often leading to a vicious cycle of anxiety and insomnia. Simple yet powerful practices like meditation, mindfulness, and deep breathing exercises can significantly mitigate stress, promoting a serene mental state conducive for sleep. Cognitive Behavioral Therapy for Insomnia (CBT-I) also holds promise, targeting the psychological aspects of sleep disturbances by identifying and altering negative thought patterns. By adopting these strategies, you can break free from the grip of stress, making way for restful nights and improved overall well-being.

Relaxation Practices

Effective relaxation practices are pivotal for managing stress and enhancing sleep quality. These practices not only help in reducing the physical symptoms of stress but can also significantly improve your emotional well-being. By incorporating relaxation techniques into your daily routine, you can create a more serene and conducive environment for sleep.

One of the simplest yet most effective relaxation practices is deep breathing. This technique focuses on slow, deep, and intentional breathing patterns that activate the body's relaxation response. Deep breathing can be done almost anywhere, making it an accessible tool to lower stress levels before bedtime. Start by sitting comfortably, closing your eyes, and taking a deep breath in through your nose, filling your lungs completely. Hold the breath for a few seconds and then exhale slowly through your mouth. Repeat this cycle several times, and you'll notice a calming effect.

Progressive muscle relaxation is another powerful method. It involves systematically tensing and then relaxing different muscle groups in your body. Beginning with your toes, tense the muscles for a few seconds before releasing them completely, moving gradually up through your legs, abdomen, chest, arms, and face. This practice helps in releasing physical tension accumulated throughout the day, paving the way for a peaceful night's sleep.

Visualization or guided imagery can also work wonders. By imagining a serene and peaceful scene, like a beach or a forest, you can mentally transport yourself to a place of calm. This practice engages your mind in a positive way, steering your thoughts away from stressors. Close your eyes and picture the details: the sound of waves, the warmth of the sun, the smell of pine trees. These rich sensory details can evoke strong feelings of relaxation.

Incorporating these relaxation practices into your nightly routine can create a buffer against the stress and anxiety often responsible for sleep disturbances. Start with one technique that resonates with you and gradually incorporate others as you become more comfortable. Remember, the effectiveness of these practices amplifies with regular use, so consistency is key.

Meditation and Mindfulness are powerful tools for improving sleep quality, especially for those struggling with insomnia and stress-related sleep issues. In our fast-paced lives, it's common to find our minds racing just as we lay down to rest. Thoughts about work, relationships, and daily responsibilities can make it difficult to wind down and drift off to sleep. This is where mindfulness and meditation come in, offering practical methods to quiet the mind and prepare it for restorative rest.

Mindfulness, simply put, is the practice of focusing on the present moment without judgment. It involves observing your thoughts and feelings from a distance rather than getting caught up in them. This practice can be particularly beneficial at bedtime. By cultivating a non-judgmental awareness of the present moment, you can interrupt the cycle of racing thoughts and anxiety that often prevent sleep. Mindfulness can be as straightforward as paying attention to your breath or the sensation of your body sinking into the mattress. Over time, these techniques can train your mind to relax more easily, enhancing your overall sleep quality.

Meditation, on the other hand, typically involves a more structured practice that aims to calm the mind and promote a state of relaxation. There are many different types of meditation, but some of the most effective for sleep include body scan meditations and guided visualizations. A body scan meditation involves mentally scanning your body from head to toe, noting any areas of tension and consciously relaxing them. This practice not only helps to release physical tension but also anchors your mind in the present moment, making it easier to drift into sleep.

One popular form of meditation for enhancing sleep is loving-kindness meditation, which involves focusing on positive thoughts and feelings towards yourself and others. By generating feelings of kindness and compassion, you can counteract the negative emotions that might be keeping you awake. Another powerful technique is guided visualization, where you listen to a recording that walks you through calming imagery, like a peaceful forest or a quiet beach. The vivid mental imagery can replace anxious thoughts, preparing your mind for a restful night's sleep.

For those new to meditation and mindfulness, starting small is key. Even just five to ten minutes of mindfulness practice before bed can make a significant difference. Apps and online resources can provide guided meditations tailored specifically to improve sleep. Consistency is crucial—making mindfulness or meditation a regular part of your bedtime routine can enhance its effectiveness over time.

In addition to improving sleep, the benefits of meditation and mindfulness extend into other areas of life. Regular practice can reduce stress, improve emotional regulation, and even enhance cognitive function. For insomniacs, these benefits can be particularly impactful. By reducing the stress and anxiety that often exacerbate sleep difficulties, mindfulness and meditation can lead to more consistent and restorative sleep patterns.

Integrating mindfulness and meditation into your nightly routine doesn't require a significant time investment but does require commitment and practice. It's about creating a habit of focusing your mind away from worries and distractions and towards peace and relaxation. As insomnia and poor sleep quality often stem from an overstimulated and stressed mind, these practices offer a natural, side-effect-free way to break the cycle.

Meditation and mindfulness should be viewed as complementary tools in your sleep improvement arsenal. They work best when combined with other healthy sleep habits, like maintaining a consistent sleep schedule, creating a restful environment, and limiting stimulating activities before bed. By addressing both the physical and mental aspects of sleep, you're setting yourself up for the best possible rest and, ultimately, a healthier, more balanced life.

If you're skeptical about the benefits, there's a wealth of scientific research supporting the positive impacts of these practices on sleep. Studies have shown that mindfulness-based interventions can improve sleep quality and reduce symptoms of insomnia. The act of meditating before bed has been linked to decreases in the time it takes to fall asleep and increases in sleep duration and quality. So, don't just take our word for it—give meditation and mindfulness a try and experience the difference for yourself.

Deep Breathing Exercises are a simple yet powerful tool for improving sleep quality and overall well-being. These exercises can be incredibly effective for both health-conscious individuals and insomniacs. Deep breathing helps calm the nervous system, reduce stress, and prepare the body for rest. Incorporating deep breathing exercises into your nightly routine can signal to your brain that it's time to wind down, making it easier to transition into a state of restful sleep.

One of the most accessible and well-known techniques is the 4-7-8 breathing method. This exercise involves inhaling deeply through your nose for a count of four, holding your breath for a count of seven, and then exhaling completely through your mouth for a count of eight. The structured pattern promotes relaxation by regulating the amount of oxygen entering and exiting your body. This shift in breathing pattern can lower your heart rate and blood pressure, making it easier to fall asleep.

Another effective technique is diaphragmatic breathing, also known as belly breathing. To practice this method, lie down in a comfortable position with one hand on your chest and the other on your abdomen. Breathe in deeply through your nose, allowing your diaphragm to fill with air and making your abdomen rise. Exhale slowly through your mouth. By focusing on using your diaphragm rather than shallow chest breaths, you enhance oxygen intake and stimulate the parasympathetic nervous system, which calms the body.

For those who find it hard to concentrate on breathing exercises alone, guided audio sessions can be beneficial. Numerous apps and online resources offer guided deep breathing practices specifically designed to aid in sleep. These guided sessions often come with soothing background music or nature sounds, adding an extra layer of relaxation. Try to follow along with a guided session for a few nights to see if it enhances your ability to relax and fall asleep.

Incorporating deep breathing exercises into your bedtime routine doesn't require significant time investment. Even just five to ten minutes of focused breathing can have marked benefits. Consistency is key; make it a nightly ritual to reap the full rewards. Pairing these exercises with other relaxing activities, such as reading a book or taking a warm bath, can amplify their effectiveness.

Deep breathing exercises can be particularly beneficial during moments of acute stress or anxiety, which often exacerbate sleep issues. If you find yourself lying in bed with racing thoughts, switch your focus to your breath. Pay attention to each inhale and exhale, letting your mind settle. This practice can divert your attention from worries and aid in quieting your mind.

It's important to create a conducive environment for these exercises. Find a quiet, comfortable spot where you won't be disturbed. Dim the lights, or use a sleep mask to block out any light sources. If possible, eliminate background noise, or use a white noise machine to create a calming atmosphere. Setting the right mood can make your deep breathing exercises more effective and enjoyable.

Moreover, deep breathing exercises can be integrated into other relaxation techniques, such as progressive muscle relaxation or meditation. By combining methods, you may find an even greater sense of calm and readiness for sleep. For instance, try practicing deep breathing while progressively tensing and relaxing different muscle groups in your body. This combination can be particularly effective in releasing physical tension and preparing your body for rest.

As you become more familiar with deep breathing techniques, you can adapt them to fit your personal preferences. Some may prefer to count inhalation and exhalation times, while others might focus on visualizing calming images or using affirmations. There's no one-size-fits-all approach; the goal is to find what best helps you relax and transition into sleep.

In summary, deep breathing exercises are a valuable addition to any sleep hygiene regimen. By regularly practicing techniques like the 4-7-8 method or diaphragmatic breathing, you can cultivate a state of relaxation that supports better sleep. Remember, the effectiveness of these exercises depends on consistency and personalization, so take the time to find what works best for you. Better sleep is within reach, one breath at a time.

Cognitive Behavioral Therapy for Insomnia (CBT-I)

One of the most effective methods for managing insomnia, especially when stress plays a major role, is Cognitive Behavioral Therapy for Insomnia (CBT-I). This structured program helps you identify and replace thoughts and behaviors that cause or worsen sleep problems with habits that promote sound sleep. Unlike sleeping pills, CBT-I addresses the underlying causes of insomnia and helps you develop long-term strategies for better sleep.

CBT-I is rooted in the idea that our thoughts, feelings, and behaviors are interconnected. Specifically, it focuses on the connections between how we think about sleep, how we feel in response to these thoughts, and how we act. This dynamic can create a vicious cycle: negative thoughts about sleep result in anxiety, which makes falling asleep harder, leading to more negative thoughts. CBT-I aims to break this cycle through a combination of cognitive and behavioral techniques.

The cognitive component of CBT-I involves recognizing and challenging unhelpful beliefs and attitudes about sleep. For instance, many insomniacs hold the belief that they must get eight hours of sleep each night to function well. CBT-I encourages you to test this belief and replace it with a more realistic attitude, such as focusing on the quality of sleep rather than the quantity. By altering your mindset, you can reduce the pressure and anxiety associated with sleep.

The behavioral aspect involves changing your habits and routines to promote better sleep. One of the primary techniques used in CBT-I is Sleep Restriction Therapy. This might sound counterintuitive, but reducing the time spent in bed can actually improve sleep quality. By strictly limiting the amount of time you allow yourself to be in bed, you can consolidate sleep to make it deeper and more restful.

Another essential behavioral strategy is Stimulus Control Therapy. This method aims to strengthen the association between bed and sleep. For example, you're advised only to go to bed when you're sleepy, to use the bed only for sleep and sex, and to leave the bedroom if you're unable to fall asleep within 20 minutes. This helps retrain your mind and body to see the bed as a place for restful sleep rather than a battleground for insomnia.

CBT-I also incorporates relaxation techniques to calm the mind and prepare the body for sleep. Practices like progressive muscle relaxation, guided imagery, and mindfulness meditation are commonly used. These techniques can lower pre-sleep arousal levels, reduce the impact of intrusive thoughts, and create a conducive state for sleep. Over time, you learn to apply these skills whenever you're feeling too anxious or stressed to sleep.

Moreover, sleep hygiene education is a critical component of CBT-I. This involves teaching you the principles of good sleep habits, such as maintaining a regular sleep schedule, creating a comfortable sleep environment, and avoiding stimulants like caffeine before bed.

Essentially, these are practical tips that support the behavioral changes CBT-I advocates.

It's worth noting that CBT-I isn't a quick fix. It requires commitment and effort, but the results are often long-lasting. Studies have shown that participants in CBT-I programs generally experience significant improvements in both the amount and quality of their sleep. Additionally, the benefits often extend beyond better sleep, contributing to reduced anxiety, improved mood, and a greater sense of well-being.

For those struggling with persistent insomnia, especially in the context of stress, CBT-I offers a comprehensive and empowering way to reclaim restful nights. By addressing both the cognitive and behavioral dimensions of sleep, this therapy helps to dismantle the barriers to good sleep, laying a foundation for healthier sleep patterns and improved overall health.

In summary, CBT-I stands out as a robust and multi-faceted approach to managing insomnia caused by stress. It empowers you to take control of your sleep patterns through a thoughtful blend of cognitive restructuring, behavioral strategies, relaxation techniques, and sleep hygiene education. Implementing these changes can lead not only to better sleep but also to an enhanced quality of life.

Chapter 11:
Technology and Sleep

In today's digital age, technology's pervasive influence can significantly disrupt our sleep patterns and quality. From the glow of smartphones to the allure of late-night TV shows, screen time can alter our natural circadian rhythm and impair melatonin production, making it harder to fall asleep. Excessive exposure to blue light, especially before bedtime, signals the brain to stay alert, effectively delaying the onset of sleep. It's crucial to implement strategies to manage screen use, such as setting device curfews and using blue light filters. Moreover, while sleep tracking devices can offer insights into our sleep habits, they should be used mindfully to avoid becoming overly fixated on perceived sleep deficiencies. Balancing technology use with good sleep hygiene practices is key to fostering restorative rest and enhancing overall well-being.

The Impact of Screen Time

In our modern world, screens are ubiquitous; they serve as our primary source of information, entertainment, and even social interaction. However, the pervasive use of screens has significant repercussions on sleep quality. Exposure to electronic devices such as smartphones, tablets, and computers, especially before bedtime, can disrupt the natural production of melatonin, the hormone that regulates sleep-wake cycles. This leads to difficulty in falling asleep and a reduction in overall sleep duration and quality. Moreover, the engaging content often found on these screens—be it social media, work emails, or streaming services—stimulates the brain, making it harder to transition to a restful state. For those striving to improve their sleep, limiting screen time in the evening is a crucial first step to reclaiming a peaceful night's rest.

Blue Light and Sleep has become a central discussion point in the realm of sleep science, particularly with the increasing prevalence of digital devices in our lives. Blue light is part of the visible light spectrum, with a wavelength between 450 and 490 nanometers, and it's emitted by many modern devices such as smartphones, tablets, computers, and LED lighting. Understanding how blue light influences sleep involves diving into how it interacts with our bodies' natural processes, notably the circadian rhythms and production of melatonin.

Circadian rhythms, our internal 24-hour clock, are highly sensitive to light, especially blue light. Exposure to blue light during the evening can disrupt these rhythms by tricking the brain into believing it's still daytime. This confusion happens because blue light inhibits the secretion of melatonin, the hormone responsible for making us feel sleepy. Normally, melatonin levels start to rise in the evening, peak during the night, and gradually decrease as morning approaches. When we expose ourselves to blue light late into the night, we delay this process, making it harder to fall asleep and stay asleep.

Studies have shown that people who use electronic devices before bed tend to take longer to fall asleep and experience poorer sleep quality. A significant research project demonstrated that individuals reading on a light-emitting device took nearly 10 minutes longer to fall asleep compared to those reading a printed book. In addition to delayed sleep onset, these individuals had shorter REM sleep phases, making them feel less rested the following day. These findings emphasize that it's not just the total amount of sleep that matters, but also the quality and structure of sleep that can be disrupted by blue light exposure.

To mitigate the negative effects of blue light, many experts recommend setting a "digital curfew." This concept involves switching off all electronic devices at least one hour before bedtime. Instead of checking emails or scrolling through social media, engage in more sleep-friendly activities such as reading a physical book, journaling, or practicing relaxation techniques. Creating a pre-sleep routine that minimizes blue light can significantly enhance sleep quality.

Another practical approach is using blue light filters on electronic devices. Most smartphones and computers offer settings that reduce blue light emission. Applications and screen protectors designed to filter out blue light are also widely available. While these tools may not entirely eliminate blue light exposure, they can help reduce its intensity and delay its impact on melatonin production. Wearing blue light-blocking glasses in the evening is another effective strategy. These glasses are designed to filter out a substantial portion of blue light, making it easier for your body to produce melatonin naturally.

It's also important to consider the lighting in your environment. Changing to warmer, dimmer lights in the evening can create a more sleep-conducive atmosphere. Many households and offices use LED lighting, which tends to emit higher levels of blue light compared to traditional incandescent bulbs. Switching to bulbs with lower color temperatures in the range of 2000-3000K can help create a more conducive environment for winding down before bed.

Some modern devices come with built-in "night modes" that automatically adjust the display's color temperature as part of the solution. These modes shift the colors of a screen to warmer hues as evening approaches, reducing blue light emission. Studies indicate that using these settings can slightly improve sleep quality, although they aren't a substitute for turning off devices entirely.

Understanding blue light's effect on sleep and implementing actionable strategies can substantially improve your sleep quality. By managing your exposure to blue light, you can support your body's natural rhythms and promote healthier sleep patterns. Whether through digital curfews, blue light filters, adjusted lighting, or blue light-blocking glasses, adopting these measures can help you achieve better rest and overall well-being.

As we move forward in this digital age, the conversation around blue light and sleep will only become more crucial. It's not just about cutting out technology entirely but making informed decisions about how we interact with it. Embracing strategies that balance our tech habits with the need for quality sleep is key to fostering a healthier relationship with both technology and sleep.

Managing Screen Use Before Bed is a critical aspect of improving sleep quality that often gets overlooked in our tech-driven society. The omnipresence of screens, from smartphones to laptops, means we're constantly exposed to blue light, which can disrupt our natural sleep patterns. So, what makes screen use before bed so detrimental to sleep, and what steps can you take to manage it effectively?

When you engage with screens before bed, you're exposed to blue light that can significantly hinder the production of melatonin, the hormone responsible for regulating sleep-wake cycles. Melatonin levels typically rise in the evening, signaling to your brain that it's time to wind down. However, blue light delays this process, tricking your brain into thinking it's still daytime and thereby prolonging wakefulness.

Moreover, the content you're consuming plays a role in affecting your sleep. Engaging in stimulating activities such as social media scrolling, gaming, or watching action-packed movies can elevate your stress levels and make it difficult to disengage mentally. As a result, it becomes harder to relax and fall asleep.

To counteract these effects, consider implementing a digital curfew. Aim to turn off all screens at least one hour before bedtime. This gives your body time to produce the melatonin needed to initiate sleep. If you find it hard to put your devices down, try setting an alarm as a reminder or use apps that can lock your devices after a certain time.

Another effective strategy is to adjust the settings on your devices to emit less blue light during the evening hours. Many modern smartphones, tablets, and computers come equipped with "night mode" or "blue light filter" settings. Enabling these features can minimize blue light exposure and reduce its impact on your sleep.

If you absolutely need to use screens before bed, consider investing in blue light-blocking glasses. These specially designed glasses can filter out blue light, making it easier for your body to maintain its natural circadian rhythms. While not a perfect solution, they can certainly help mitigate the effects of late-night screen use.

Creating a calming pre-sleep routine can also work wonders for your sleep quality. Instead of reaching for your phone, engage in relaxing activities like reading a physical book, taking a warm bath, or practicing mindfulness exercises. These activities can help signal to your brain that it's time to prepare for sleep.

Content selection also matters. Opt for soothing, less stimulating content if you must engage in screen time. Meditation apps, audiobooks, and podcasts with calming narratives are excellent alternatives to social media or action movies. These can ease you into a more relaxed state, conducive to sleep.

Lastly, be mindful of your bedroom environment. Keep the room cool, dark, and screen-free as much as possible. Reserve your bed for sleep and intimate activities only, to strengthen the mental association between being in bed and sleeping. This behavioral conditioning can significantly improve your sleep quality over time.

In summary, managing screen use before bed is crucial for enhancing sleep quality. By setting a digital curfew, adjusting device settings, using blue light-blocking tools, and creating a relaxing pre-sleep routine, you can minimize the negative impacts of screens on your sleep and enjoy a more restful night.

Sleep Tracking Devices

Sleep tracking devices have revolutionized how we understand and approach sleep. These gadgets, ranging from wearable smartwatches to advanced medical instruments, offer a closer look at our nightly habits and overall sleep health. They work by monitoring various physiological parameters like heart rate, body movements, and even oxygen levels. By doing so, they provide valuable data insights that help in identifying patterns, improving sleep quality, and diagnosing potential sleep disorders.

The technology behind sleep trackers varies widely. Basic models use accelerometers to detect movements, which helps estimate sleep stages and waking periods. More advanced devices integrate heart rate monitors and even electrocardiograms (ECGs) to provide a comprehensive analysis of sleep cycles. Additionally, some cutting-edge models use sensors to measure breathing irregularities, which can be crucial in identifying conditions like sleep apnea.

One of the main benefits of sleep tracking devices is their ability to highlight trends over time. Instead of guessing or using subjective measures, you can rely on concrete data to understand your sleep patterns. Imagine seeing a week-long trend of reduced deep sleep correlated with late-night screen time. Understanding these patterns empowers you to make informed decisions about your sleep hygiene and daily habits.

Despite their benefits, these devices aren't without limitations. For one, the accuracy can vary significantly between different brands and models. While they generally do a good job of logging sleep duration and general patterns, they may not always accurately capture the intricacies of sleep stages or differentiate between light and deep sleep with medical-grade precision.

It's also worth mentioning the psychological effects. Constant tracking can create anxiety around sleep, sometimes paradoxically worsening sleep quality. It may lead to an obsession with data points rather than focusing on creating a genuinely restful environment. Thus, while tracking can be a powerful tool, it should be used wisely and not be the sole focus of your sleep improvement efforts.

For those managing specific sleep disorders, these devices can offer significant support. Insights from sleep trackers can help healthcare professionals fine-tune treatment plans. For example, individuals suffering from insomnia can use the data to identify triggers and work with their doctors on targeted interventions. People with diagnosed sleep apnea can monitor the efficacy of their treatments, such as CPAP therapy.

The convenience of modern sleep trackers cannot be understated. With smartphone integrations, you can view detailed reports, set sleep goals, and even get personalized recommendations to improve your nightly rest. These user-friendly interfaces make it easier for everyone—from tech novices to health enthusiasts—to engage proactively with their sleep health.

In summary, sleep tracking devices offer a wealth of information and potential benefits for improving sleep quality. However, their data should be a guide, not a source of stress. Coupling these insights with sound sleep hygiene practices and professional advice will give you the best chance at achieving restful, restorative sleep.

Chapter 12:
Special Considerations

When considering sleep, one size does not fit all. Each stage of life presents unique challenges and requirements for optimal sleep health. From the fluctuating sleep patterns of children and adolescents, which are often disrupted by hormonal changes and social pressures, to the altered sleep architecture of older adults, who may face issues like advanced sleep phase syndrome, each demographic group requires tailored advice and interventions. Gender differences also play a crucial role, as hormonal cycles can profoundly affect women's sleep quality. Recognizing and addressing these nuanced variations ensures that sleep strategies are not only effective but also inclusive, providing everyone with the best possible chance to enjoy restorative sleep. Understanding these special considerations helps us create more personalized and compassionate sleep solutions, ultimately fostering better overall well-being.

Sleep Across the Lifespan

Sleep needs and patterns change significantly across the lifespan. From infancy to old age, our requirements for restorative slumber evolve, influenced by a myriad of physiological, psychological, and social factors. Understanding these shifts is crucial for tailoring sleep hygiene practices and improving overall well-being.

During infancy, babies spend the majority of their time sleeping. Their sleep cycles are shorter, and they oscillate quickly between active

(REM) and quiet (non-REM) sleep. This is fundamental for brain development and physical growth. As children transition into adolescence, their internal body clocks naturally shift, causing a biologically driven preference for later bedtimes and wake-up times. This phenomenon, known as "sleep phase delay," can clash with early school start times, contributing to a widespread issue of sleep deprivation among teenagers.

Adulthood brings relatively stable sleep patterns, yet life's pressures—from career demands to parenting challenges—can disrupt this stability. Adults typically need between 7 to 9 hours of sleep per night, but quality is just as essential as quantity. Stress, health issues, and poor lifestyle choices often lead to fragmented sleep, which can affect mental acuity, emotional balance, and physical health.

As we reach our senior years, sleep architecture changes again. Older adults experience a natural decrease in deep sleep (stages 3 and 4 of non-REM sleep) and may have more fragmented sleep with frequent awakenings. This can be due to a decline in the production of melatonin, the hormone responsible for regulating sleep-wake cycles, or the presence of chronic conditions such as arthritis or sleep apnea. A common misconception is that older individuals need less sleep; however, they still require about 7 to 8 hours of restful sleep to maintain health and cognitive function.

By better understanding these age-related changes in sleep, we can implement targeted strategies to enhance sleep quality at every stage of life. From advocating for later school start times for teenagers to encouraging relaxation techniques for older adults, recognizing and adapting to the body's evolving sleep needs is essential for longevity and well-being.

Sleep in Children and Adolescents plays a pivotal role in their overall health and development. Unlike adults, children and adolescents require more sleep to support their rapid physical and cognitive growth. Numerous studies highlight that ensuring adequate sleep during these formative years can significantly impact their academic performance, emotional well-being, and even their long-term health.

The amount of sleep needed varies by age group. Young children, particularly those between the ages of 3 to 5, typically require 10 to 13 hours of sleep per night. As children grow older, their sleep requirement slightly decreases, with school-aged children (6 to 12 years) needing 9 to 12 hours per night. Adolescents, who are often juggling school, extracurricular activities, and social lives, should aim for 8 to 10 hours of sleep each night. However, the reality is starkly different; studies indicate that a significant percentage of adolescents are sleep-deprived.

One major culprit of sleep deprivation in adolescents is the increasing academic pressure and corresponding homework load. Coupled with extracurricular demands and social activities, it becomes challenging for teenagers to prioritize sleep. This pressure often results in late-night study sessions, which can disrupt their natural circadian rhythm, leading to irregular sleep patterns.

Another critical factor is the pervasive presence of technology. Smartphones, tablets, and computers have become almost ubiquitous in the lives of young people. The blue light emitted by these devices can interfere with the production of melatonin, a hormone essential for sleep regulation. As a result, children and adolescents who engage with screens before bedtime often experience delays in falling asleep.

It's not just about the duration, though; the quality of sleep is equally important. Poor sleep quality in children and adolescents can manifest as frequent awakenings, difficulty in maintaining sleep, or even conditions like sleep apnea. These disruptions can have cumulative adverse effects on their physical health, emotional stability, and cognitive functioning.

Parents and caregivers play a crucial role in fostering good sleep hygiene among children and adolescents. Establishing a consistent bedtime routine can have a profound impact. This routine might include calming activities such as reading a book, taking a warm bath, or practicing relaxation exercises. Consistency is key; going to bed and waking up at the same time every day, even on weekends, helps reinforce the body's natural sleep-wake cycle.

Creating a sleep-friendly environment is another imperative. The child's bedroom should be quiet, dark, and cool, which are conducive conditions for sleep. Limiting exposure to noise and light can significantly enhance sleep quality. Additionally, it's advisable to remove distractions such as TVs and computers from the bedroom to promote an environment solely dedicated to rest.

Nutrition also plays a part in sleep regulation for kids and teens. Consuming caffeine, found in sodas, energy drinks, and chocolate, especially in the late afternoon or evening, can adversely affect their ability to fall asleep. A balanced diet that includes foods rich in tryptophan, magnesium, and calcium can support better sleep.

The role of physical activity in promoting good sleep cannot be understated. Regular physical exercise has been shown to help children and adolescents fall asleep faster and enjoy a deeper sleep. However, it's important to note the timing – vigorous activity close to bedtime can have the opposite effect, making it harder for them to wind down.

Mental health is intricately linked with sleep in children and adolescents. Stress, anxiety, and depression can severely disrupt sleep patterns. In such cases, utilizing relaxation techniques or seeking professional help, such as cognitive behavioral therapy, can be beneficial. Teaching kids mindfulness and stress-management techniques early can provide them with tools to navigate the pressures of their daily lives, positively impacting their sleep as well.

Peer pressure and social expectations can also influence sleep habits. Adolescents, in particular, may feel compelled to stay up late to fit in socially, or they might experience 'social jetlag' – a mismatch between their biological clock and social obligations.

The impact of sleep on academic performance is profound. Studies have repeatedly shown that students with adequate sleep have better attention spans, improved memory retention, and higher cognitive abilities, which translate into better academic outcomes. Conversely, sleep deprivation can lead to difficulties concentrating, lower problem-solving abilities, and increased feelings of fatigue and irritability, all of which can hinder academic progress.

Lastly, it's crucial to raise awareness among educators and policymakers about the significance of sleep for young people. Some schools are taking progressive steps by delaying school start times to align better with the natural sleep patterns of adolescents, acknowledging the science that supports these changes.

In conclusion, sleep in children and adolescents is not just a necessity but a fundamental aspect of their health and development. By understanding the unique sleep needs of this age group and implementing practical strategies to promote better sleep hygiene, we can help our younger generation achieve their full potential both during sleep and waking hours.

Sleep in Older Adults undergoes significant changes that can impact both the quality and quantity of rest. This is not merely a facet of aging, but a convergence of physiological, psychological, and lifestyle alterations. Understanding these shifts is crucial for anyone keen on enhancing their sleep quality and overall well-being.

As we age, our sleep architecture—a blueprint of our nightly sleep patterns—modifies. Older adults typically experience a decrease in deep sleep, also known as slow-wave sleep, and a corresponding increase in lighter stages of sleep. This transition can make sleep feel less refreshing, even if the total amount of sleep remains the same. Studies have shown a reduction in the duration of REM sleep, which is crucial for cognitive function and emotional health. The decreased presence of REM and deep sleep stages can contribute to older adults frequently feeling tired or unrefreshed upon waking.

One of the most profound changes in sleep patterns in older adults is a shift in circadian rhythms. Often referred to as a "phase advance," this shift causes older adults to feel sleepy earlier in the evening and wake up earlier in the morning. While this might seem a benign adjustment, it can lead to issues if the individual's lifestyle or social engagements don't align with this new rhythm. Maintaining a consistent sleep schedule becomes essential to counteracting the effects of such phase advances.

The production of melatonin—a hormone responsible for regulating sleep-wake cycles—also declines with age. The pineal gland, which secretes melatonin, becomes less active. This lack of melatonin can make initiating and maintaining sleep more difficult. As a result, older individuals may find it harder to adapt to irregular sleep schedules or night-time disruptions.

Aside from biological changes, older adults often face an increase in chronic health conditions, such as arthritis and cardiovascular issues, which can hinder a good night's sleep. Persistent pain and discomfort from these ailments can lead to frequent awakenings and a general fragmentation of sleep. Addressing these pain points through appropriate medical interventions, exercise, and a comfortable sleeping environment can help improve sleep quality.

Medication use in older adults is another component to consider. Various prescriptions for conditions like high blood pressure or depression can interfere with sleep. Beta-blockers, commonly used for heart conditions, can inhibit melatonin production, while some antidepressants might alter REM sleep. It's crucial for older adults and their healthcare providers to review medications frequently and consider possible sleep-related side effects.

Psychological factors including loneliness, anxiety, and depression can also greatly affect sleep in older adulthood. Cognitive Behavioral Therapy for Insomnia (CBT-I) has shown promising results in treating sleep disturbances related to these mental health issues. CBT-I focuses on changing sleep habits and misconceptions that perpetuate sleep difficulties, offering a structured program to improve sleep quality.

Traditional sleep hygiene techniques become even more valuable as we age. Maintaining a bedroom environment that promotes sleep— such as limiting noise, ensuring the room is dark, and maintaining a comfortable temperature—can make a world of difference. Additionally, engaging in relaxing activities before bed, like reading or gentle stretching, can help signal to the body that it's time to wind down.

Physical activity during the day can also have profound impacts on sleep for older adults. Regular exercise, tailored to an individual's capabilities and health status, can help in falling asleep faster and enjoying a deeper sleep. However, it's essential to time workouts appropriately, as exercising too close to bedtime can have the opposite effect, increasing alertness and making it harder to fall asleep.

Daytime napping is a common practice among older adults, but it's important to regulate these naps to avoid disrupting the nightly sleep pattern. Short naps (20-30 minutes) in the early afternoon can be rejuvenating without interfering with nighttime sleep. Long or late afternoon naps, however, can make it tougher to fall asleep at night.

Nutritional considerations also play a role in the sleep health of older adults. Certain foods and drinks, such as caffeine and alcohol, can have pronounced effects on sleep. While a glass of wine might help someone fall asleep initially, alcohol disrupts the second half of the sleep cycle, causing restlessness and early waking. Conversely, a diet rich in fruits, vegetables, and whole grains can promote better sleep hygiene by providing the necessary vitamins and minerals that contribute to a balanced sleep cycle.

Managing sleep hygiene in older adults often requires a more comprehensive and integrated approach compared to younger individuals. By acknowledging and addressing physiological, psychological, and lifestyle factors, improvements in sleep quality and overall well-being are entirely achievable. After all, good sleep is not a luxury, but a cornerstone of healthy aging.

Gender and Sleep Differences are a nuanced and fascinating area of study in sleep science. Men and women experience sleep differently due to a myriad of physiological, hormonal, and psychosocial factors. These differences can impact not only the quality of sleep but also the overall well-being of each gender. Although some variations are subtle, others are more pronounced and can be influenced by life stages such as puberty, pregnancy, and menopause.

One of the core differences lies in sleep architecture. Research has shown that women generally have longer sleep durations and more slow-wave sleep (deep sleep) compared to men. This might sound like an advantage, but women are also more prone to experiencing sleep disorders, particularly insomnia. Hormones play a pivotal role here. For instance, the menstrual cycle involves fluctuations in estrogen and progesterone, which can lead to sleep disturbances like night sweats and increased restless nights.

During pregnancy, sleep can become even more elusive for women. Factors such as hormonal changes, physical discomfort, and increased urination can severely disrupt sleep. The third trimester is often the most challenging phase, with many women reporting significant increases in sleep fragmentation. Adding to this complexity, postpartum sleep can be erratic, impacting new mothers' mood and cognitive function.

Men, on the other hand, are more prone to sleep apnea, a condition characterized by repeated interruptions in breathing during sleep. This difference is primarily due to variations in upper airway anatomy and fat distribution patterns between men and women. Men also tend to have higher rates of snoring and nighttime awakenings, which can be exacerbated by lifestyle factors such as alcohol consumption and obesity.

Beyond biological factors, societal roles and stressors contribute significantly to gender disparities in sleep. Women often juggle multiple responsibilities, including work, childcare, and household chores, which can increase stress levels and impact sleep quality. Interestingly, studies indicate that women are more likely to seek help for sleep problems than men, possibly because men are less inclined to discuss or acknowledge their sleep issues. This difference in health-seeking behavior can result in variations in the diagnosis and treatment of sleep disorders between genders.

As women age, particularly during the menopausal transition, hormonal changes become even more pronounced. The decline in estrogen can lead to symptoms such as hot flashes and mood swings, further complicating sleep. Men, although they also experience hormonal changes with age (like a gradual decrease in testosterone), do not appear to have their sleep as severely impacted as women during this phase of life.

Understanding these gender-specific differences in sleep is crucial for tailoring treatment and improving sleep hygiene. For women, addressing hormonal issues and managing stress through relaxation techniques could be particularly beneficial. Cognitive Behavioral Therapy for Insomnia (CBT-I) and mindfulness practices can offer effective strategies for mitigating sleep disruptions caused by stress and hormonal fluctuations.

For men, focusing on lifestyle modifications such as weight management, reducing alcohol intake, and undergoing routine screenings for sleep apnea can be crucial steps. Additionally, since men are less likely to seek help, raising awareness about the importance of sleep and encouraging open discussions about sleep health can make a significant difference.

These insights into gender and sleep differences also have practical implications for healthcare providers. Tailoring advice and interventions based on an individual's gender can lead to more effective treatments and improved sleep quality. Personalized sleep solutions can help bridge the gap and ensure that both men and women achieve optimal rest and overall well-being.

In conclusion, the interplay between gender and sleep is complex and multidimensional. Through understanding these differences, we can take actionable steps to address the unique sleep challenges faced by men and women. Enhanced awareness and targeted interventions can pave the way for better sleep and a healthier, more fulfilling life.

Conclusion

As we reach the end of our exploration into the world of sleep, it becomes clear that sleep is not just a passive state of rest, but a dynamic process essential for our well-being. Throughout this book, we've delved into the intricacies of sleep—the physiological mechanisms, psychological implications, and the myriad of health benefits it bestows. Our journey has underscored one undeniable truth: prioritizing sleep is one of the most profound acts of self-care we can perform.

Understanding sleep starts with grasping its fundamental architecture, from the circadian rhythms that dictate our sleep-wake cycles to the distinct stages of sleep that play unique roles in restoration and repair. Equipped with this knowledge, we can start making informed decisions to enhance our sleep quality. Remember, sleep isn't merely about lying down and closing your eyes; it's about creating an environment and routine that facilitates the deep, restorative rest our bodies crave.

The interplay between sleep and our mental health cannot be overstated. Chronic sleep deprivation impacts not just our mood but also our cognitive functions and emotional resilience. Addressing issues like insomnia or sleep apnea, whether through lifestyle changes, therapeutic interventions, or medical treatment, can dramatically improve our quality of life. And while the challenges may vary—stress, anxiety, physical discomfort—tailored strategies can make a world of difference.

From diet and exercise to managing stress and minimizing screen time, our daily habits either build a bridge to restorative sleep or erect barriers against it. Nutritional choices, for instance, play a pivotal role. Foods rich in tryptophan, magnesium, and melatonin can ease your path to sleep, whereas caffeine and alcohol often disrupt it. Similarly, regular physical activity needs to be timed wisely to avoid interference with your sleep patterns.

The societal changes in recent decades, particularly our relationship with technology, present new challenges and opportunities. Understanding the impact of blue light and developing habits around screen use are vital in navigating these modern obstacles. While technological advancements have brought us sleep tracking devices that can provide valuable insights, they also remind us to strike a balance between connectivity and restorative downtime.

Ultimately, enhancing sleep quality demands a personalized approach—one that considers individual differences in lifestyle, physical health, and emotional well-being. Whether you're young or old, male or female, each life stage presents unique sleep challenges and solutions. Cultivating good sleep hygiene, leveraging nutritional insights, and adopting stress-management techniques can collectively pave the way to better sleep.

In conclusion, the key takeaway is that sleep is not a luxury but a necessity, deeply intertwined with almost every aspect of our health. By making informed choices and fostering habits that respect the natural rhythms of our bodies, we take significant steps toward enhancing our overall well-being. Here's to transforming knowledge into action and reclaiming the restorative power of good sleep for a healthier, happier life.

Appendix

This appendix serves as a consolidation of practical information and supplementary resources to help optimize your sleep quality and overall well-being. You will find detailed charts, worksheets, and additional reading materials that act as tools to put the book's lessons into action. Whether you're tracking your sleep patterns or looking for further reading on specific topics, this section aims to provide clear, actionable insights in a user-friendly manner. By incorporating the guidance from this appendix, you can make informed decisions to tailor your sleep environment, routines, and lifestyle choices, all geared towards fostering deeper, more restorative sleep.

Glossary
of Sleep Terms

This glossary provides definitions and explanations of key terms related to sleep science and the regulation, disorders, and benefits of sleep. It's designed to serve as a quick reference for health-conscious individuals and insomniacs looking to deepen their understanding of sleep.

Adenosine

A neurotransmitter that builds up in your brain during the day, promoting a feeling of sleepiness as it accumulates.

Alpha Waves

Type of brain waves detected either by EEG or MEG, prominent when you are awake but relaxed and calm, typically with eyes closed.

Apnea

A sleep disorder where you experience pauses in breathing or shallow breaths while you sleep. Commonly associated with Sleep Apnea.

BRAC (Basic Rest-Activity Cycle)

A 90-minute cycle of activity and rest that continues both when awake and asleep, impacting the sleep cycle.

Chronotype

A person's natural predisposition with regard to the times of day they prefer to sleep or when they feel most alert and energetic.

Circadian Rhythm

Internal body clock regulating the approximately 24-hour cycle of biological processes, including sleep-wake cycles.

Delta Waves

High amplitude brain waves associated with the deep sleep stages (NREM Stage 3), playing a crucial role in restorative sleep.

Dopamine

A neurotransmitter that influences mood, motivation, and sleep regulation by interacting with other sleep-inducing chemicals such as melatonin.

Insomnia

A common sleep disorder that makes it hard to fall asleep, stay asleep, or cause you to wake up too early and not be able to get back to sleep.

Melatonin

A hormone produced by the pineal gland in the brain that regulates sleep-wake cycles. It is often called the sleep hormone.

NREM (Non-Rapid Eye Movement) Sleep

The portion of sleep that encompasses the lighter stages 1 and 2, and deep slow-wave sleep, or stages 3 and 4.

Obstructive Sleep Apnea (OSA)

A serious sleep disorder where the throat muscles intermittently relax and block the airway during sleep, causing breathing to repeatedly stop and start.

Polysomnography

A comprehensive recording of the biophysiological changes that occur during sleep, often referred to as a sleep study.

REM (Rapid Eye Movement) Sleep

The sleep phase characterized by rapid eye movements, more dreaming and bodily movement, and faster pulse and breathing. It is critical for cognitive functions like memory consolidation.

Serotonin

A neurotransmitter that contributes to feelings of well-being and happiness, also playing a role in regulating sleep and wakefulness.

Sleep Cycle

The progression through a series of stages—NREM to REM sleep—that cycles multiple times throughout a typical night's sleep, lasting about 90-120 minutes each.

Sleep Hygiene

Habits and practices that are conducive to sleeping well on a regular basis, such as maintaining a regular sleep schedule and creating a restful environment.

Sleep Inertia

The grogginess and cognitive impairment one feels upon waking from sleep, especially from deep sleep stages.

Sleep Latency

The amount of time it takes to transition from being fully awake to sleeping, typically measured in a sleep study.

Sleep Spindles

Sudden bursts of oscillatory brain activity visible on an EEG that occur during NREM Stage 2 sleep, believed to be involved in memory consolidation and synaptic plasticity.

Sleep-Wake Homeostasis

A regulatory process that generates an increasing drive to sleep as the length of wakefulness increases, and subsequently, reduces this drive during sleep.

Suprachiasmatic Nucleus (SCN)

A cluster of cells in the hypothalamus in the brain that governs the circadian rhythms, acting as the master clock.

Theta Waves

Brain waves that are most prevalent in Stages 1 and 2 of NREM sleep, playing a role in the transition between wakefulness and sleep.

Ultradian Rhythm

Recurrent periods or cycles repeated throughout a 24-hour circadian day, such as the cycles of the various sleep stages.

White Noise

A consistent, unobtrusive background sound that can help mask other noises, potentially aiding in improved sleep.

We hope this glossary helps you better understand the terminology used throughout the book and in your journey towards better sleep health.

Additional Resources and Recommended Reading

To further your understanding and enhance your sleep quality, exploring additional resources can be incredibly beneficial. Here's a curated list of books, articles, and websites that offer valuable insights and practical tips related to the topics covered in this glossary.

"Why We Sleep: Unlocking the Power of Sleep and Dreams"

by Matthew Walker

An enlightening read that dives deep into the importance of sleep, the science behind it, and practical advice for better sleep health.

"The Sleep Revolution: Transforming Your Life, One Night at a Time"

by Arianna Huffington

Offers a comprehensive look at sleep's role in our lives and practical tips for achieving better rest.

The National Sleep Foundation - *www.sleepfoundation.org*

A reliable website filled with research, information, and practical guides on sleep topics ranging from sleep disorders to sleep hygiene.

"Say Good Night to Insomnia"

by Gregg D. Jacobs

Based on cognitive behavioral therapy principles, this book provides a structured program for overcoming insomnia without medication.

***PubMed* - www.ncbi.nlm.nih.gov/pubmed**

An excellent resource for accessing scientific studies and articles related to sleep research.

"No More Sleepless Nights"

by Peter Hauri and Shirley Linde

This classic book offers practical and scientifically-backed methods to improve sleep and tackle insomnia.

Sleep Research Society - **www.sleepresearchsociety.org**

A professional society dedicated to promoting sleep research and sleep medicine.

These resources offer a blend of scientific insights, practical strategies, and supportive guidance tailored to both health-conscious individuals and those struggling with sleep issues. By delving into these readings, you'll gain a deeper understanding of sleep and how to optimize it for better health and well-being.